AGRICULTURAL ENERGETICS

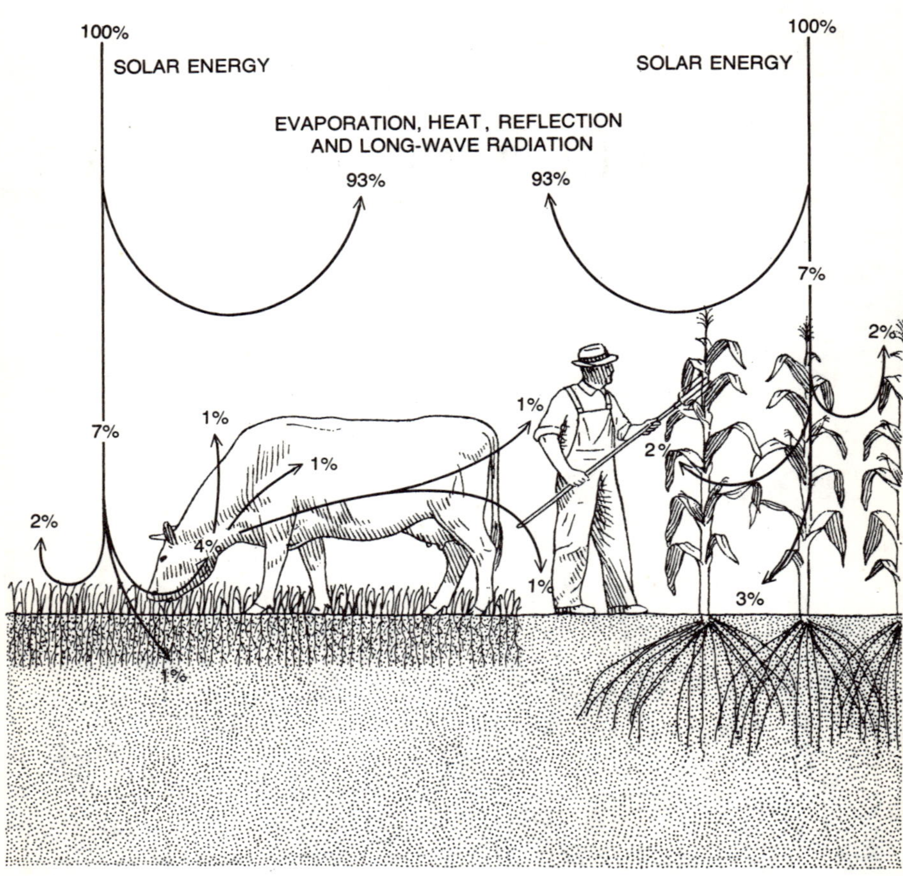
100%
SOLAR ENERGY
100%
SOLAR ENERGY
EVAPORATION, HEAT, REFLECTION
AND LONG-WAVE RADIATION
93%
93%
7%
7%
2%
1%
1%
4%
1%
1%
1%
2%
2%
3%

AGRICULTURAL ENERGETICS

Richard C. Fluck, Ph.D.,
Professor

C. Direlle Baird, Ph.D.,
Assoc. Professor

Agricultural Engineering Department
Institute of Food and Agricultural Science
University of Florida, Gainesville

AVI PUBLISHING COMPANY, INC.
Westport, Connecticut

Library of Congress Cataloging in Publication Data

Fluck, Richard C
Agricultural energetics.

Includes index.
1. Agriculture—Energy consumption. 2. Agriculture—Energy conservation. 3. Food industry and trade—Energy consumption. 4. Food industry and trade—Energy conservation. 5. Bioenergetics. I. Baird, Carl Direlle, joint author. II. Title.
S494.5.E5F58 631 79-24260
ISBN 0-87055-346-1

Printed in the United States of America

Preface

Energy's importance to agriculture and to the food chain is universally accepted. Yet there is much confusion and misunderstanding over the functioning of energy in agricultural systems and much controversy over how agricultural systems should be modified in response to limited supplies of higher priced energy. Some of this confusion results from the fact that agriculture is one of only a few industries which directly utilizes solar energy, or from the fact that agricultural products have caloric energy.

Agricultural Energetics describes our conception of how energy causes agricultural and food systems to function. The various flows of energy are identified and quantified, enabling one to make accurate and useful energy analyses of agricultural systems.

We believe this book will serve usefully as a textbook for courses on agricultural energetics or for courses including agricultural energetics as a portion of the subject matter. This book should also prove highly useful as a reference for scientists of various agricultural disciplines, policy makers, planners, energy analysts and energy extension service personnel.

Our thanks are directed toward Ernie Smerdon who in 1973 encouraged us to develop and teach a course on energy and food production. The resultant interaction with enthusiastic students from many disciplines has contributed signficantly to the breadth and depth of the subject matter. *Agricultural Energetics* draws heavily from the class notes we developed over the five years we have taught the course. To write this book was an opportunity to further organize and sharpen our concepts, which we greatly appreciate.

The authors would also like to express their thanks to Dr. D.K. Tressler, Dr. N. W. Desrosier and Ms. Lisa E. Melilli of the AVI Publishing Company for encouragement and assistance in bringing this book into being.

No book is ever finished according to the authors' perception of its needs for further additions and refinement. This one is no exception. However, we are well aware that the areas of energy analysis in general and agricultural energetics in particular are rapidly advancing. Therefore, this being the First Edition, we would greatly appreciate communications from readers for suggestions or recommendations on how to improve it and also to call to our attention errors which may be corrected in the next printing.

RICHARD C. FLUCK
C. DIRELLE BAIRD

July, 1979

Contents

1

Fundamentals of Energy and Agriculture

Energy is both the fuel and the feedstock for agriculture. Primary energy is the fuel; solar energy is the feedstock material. Both are essential in industrialized agriculture.

Concern for the efficient and proper use of energy in industrialized agriculture is evidenced from many quarters, including many nonagricultural sources. Widely differing proposals have been made for changes in agricultural policy in response to these energy-related concerns. Controversy exists over the correctness and feasibility of many of these proposals. It is of extreme importance, therefore, that the energetic relationships of agricultural systems be known and understood. We must be able to correctly identify and measure the flows of energy in the systems comprising crop and livestock production, food processing, distribution and preparation. We call such activities agricultural energetics.

It is painfully evident that energy costs have risen sharply since 1973. The spector of future energy shortages dwarfs the long and slow-moving service station lines of the winter of 1973–74. Energy's increasing costs and decreasing availability affects agriculture, which in turn affects everyone. Though trite, it is seemingly forgotten by some that food is a basic necessity for life. Many of our other activities are not so important as those of our food system.

The emphasis here is on the consumption and depletion of nonrenewable energy resources. The ultimate bottom line the authors attempt to reach is the amount of primary energy required for the production of an agricultural product. The required primary energy is conventionally measured as output of the energy producing sectors of the economy, i.e., coal mining, crude oil and natural gas production, and their equivalents in hydroelectric and nuclear power.

The authors attempt to be exact and precise in use of terms associated with energy analysis. Many terms have, however, come into common usage without proper definitions. And, as with any infant discipline, there is confusion and conflict over definitions and usage. The authors have provided a list of definitions and sources in Appendix A.1.

Many different energy units are in common usage around the world. This book uses the metric (SI) system, and its energy unit, the Joule. Most energy quantities are expressed in the more convenient quantity megaJoule (MJ) or 10^6 Joules (J). Appendix Table A.2. gives conversions among various energy units.

ENERGY FUNDAMENTALS

Several definitions are basic to an understanding of the nature of energy:

(1) *Force* (push or pull): A vector quantity (a quantity having both magnitude and direction) tending to produce change in the motion of objects. Measured in units such as pound force, kilogram force, dyne and Newton.
(2) *Work:* Product of force and distance. Measured in units such as foot-pound and Newton-meter (Joule).
(3) *Energy:* Capacity to perform work and is equivalent to work. Forms of energy include heat or thermal, chemical, electrical, radiant and nuclear energy, etc. Measured in units such as British thermal unit (Btu), kilocalorie (kcal), Joule (J), kilowatt-hour (kWh), footpound (ft lb) and horsepower-hour (hp · h).
(4) *Power:* Time rate at which work is done or energy is expended or generated. Measured in units such as watt ($J \cdot s^{-1}$) and horsepower (550 ft lb $\cdot s^{-1}$), where s = second.

The two laws which govern energy conversions are the first and second laws of thermodynamics which are:

(1) *Conservation of energy:* Energy cannot be created or destroyed, but it can be changed from one form to another.
(2) *Entropy law, irreversibility of all natural processes:* It is impossible to construct an engine that, operating in a cycle, will produce no effect other than the extraction of heat from a reservoir and the performance of an equivalent amount of work. In other words, some heat must be rejected to another (lower temperature)

reservoir. Another statement of the second law is that heat flows by itself only from the hotter to the colder body or reservoir, never in reverse. Entropy is a quantity closely associated with the second law; change in entropy is defined as the quotient of the energy discarded during a process and the lowest temperature available for heat rejection. Entropy remains constant for a reversible (ideal, and, so far as is known, nonexistent) process and increases for every irreversible process. The total entropy of the universe is relentlessly increasing, although the entropy of individual objects, organisms and systems may be caused to decrease.

Energy in one form may be converted to another form. For instance, steam may be converted to electrical power by a generator. No single conversion will result in all the input energy being converted to the desired form of output energy. Efficiency measures the ratio of output to input and, in doing so, quantifies the amount lost, usually as waste heat.

$$\text{Efficiency} = \frac{\text{Useful Work}}{\text{Input Energy}} = \frac{\text{Input Energy-Waste Heat}}{\text{Input Energy}}$$

Conversions of chemical energy to mechanical work or electrical power are typically 20–40% efficient. Conversions of chemical energy to heat are typically 60–80% efficient. Conversions of electric power to mechanical are usually 60–90% efficient. Conversions of chemical to mechanical by animals are typically 3–10% efficient.

Conservation denotes the efficient use of energy, i.e., in conversions of high efficiency and elimination of waste energy. Fluorescent lamps produce the same amount of light for less electric power than incandescent lamps; therefore, fluorescent lamps are more efficient and conserve more energy than incandescent lamps.

Energy exists in many forms. One useful energy classification scheme is renewable and nonrenewable.

Nonrenewable	*Renewable*
Fossil Fuels	Solar
Natural gas	Sunlight
Oil	Wind
Coal	Hydroelectric
Peat	Photosynthesis
Nuclear	Tidal
	Geothermal

Nonrenewable fuels are often termed *stock* and renewable fuels *flow*.

BIBLIOGRAPHY

HEICHEL, G.H. 1976. Agricultural production and energy resources. Amer. Sci. *64,* 64–72.

HUBBERT, M.K. 1971. The energy resources of the earth. Sci. Amer. *224* (3) 60–70.

SUMMERS, C.M. 1971. The conversion of energy. Sci. Amer. *224* (3) 148–160.

2

Energy in Nonindustrial Food and Production Systems

Energy is necessary for food production. Solar energy is the source of energy for food in natural ecosystems. It is supplemented by energy from other sources—primarily fossil fuels, such as crude oil, natural gas and coal—in some agricultural systems. Fossil fuels are the predominant forms of energy in most industrialized agricultural systems and are the predominant energy sources in subsequent portions of the food system in industrialized societies.

Food contains energy which is partially released to the consumer. Nonfood agricultural products, fibers such as cotton and wool and ornamentals such as foliage and flowers, also contain energy but ordinarily this is not an important characteristic of such products.

NATURAL ECOSYSTEMS

Natural ecosystems run on energy in a self-regulating manner, solar energy being the fuel for such systems. Photosynthesis is the chemical conversion of solar energy by plants into food products or carbohydrate forms. Plant parts, serving as energy sources, are eaten by herbivores or decomposed by microorganisms. Carnivores consume other animal life for their energy sources. Therefore, all forms of life in a natural ecosystem depend ultimately on the sun for the energy necessary to sustain life.

The rate of incoming solar energy at the outer limits of the earth's atmosphere is indicated by the solar constant, $1.395 \times 10^{-3}\,MJ \cdot m^{-2} \cdot s^{-1}$. After depletion by hours of darkness, atmospheric absorption, cloudiness, etc., the solar energy (insolation) on the earth varies from an average of 0.09×10^{-3} MJ $\cdot$ $m^{-2} \cdot s^{-1}$ at polar regions to 0.29×10^{-3} MJ $\cdot$ $m^{-2} \cdot s^{-1}$ in desert areas. In the summer, a typical United States location will receive a total of 20 to 30 MJ $\cdot$ m^{-2} in the course of a day. The earth, during an average day of the year, receives about 1.49×10^{16} MJ $\cdot$ d^{-1}, where d = day.

What happens to the solar energy coming into the earth's atmosphere? About 30% is directly reflected as short wavelength radiation. Another 47% is converted to heat by the atmosphere or by the earth and reradiated to outer space as long wavelength radiation. Another 23% evaporates water and thus drives the earth's hydrologic cycle. Only 0.2% drives the earth's weather system through wind and waves. Only 0.02% or about 3.5×10^{12} MJ $\cdot$ d^{-1} is used in photosynthesis by plants.

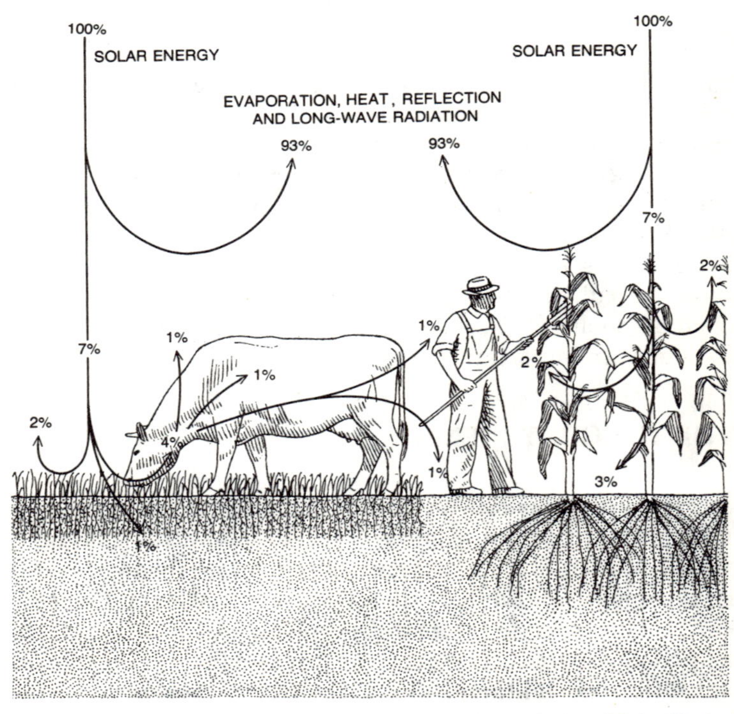

From Loomis (1976) with permission of Scientific American

FIG. 2.1. PARTITIONING OF ENERGY INVOLVED IN PRODUCING CROPS SUCH AS CORN AND GRASS

Photosynthesis is the fundamental photochemical reaction of life. As photosynthesis occurs in plants, CO_2, H_2O and radiant energy are combined to form O_2 and carbohydrates. Only wavelengths of light centering on blue and red, which consist of about 25% of the incoming solar radiation, are utilized in photosynthesis. Solar energy converted by photosynthesis is contained and stored in the carbohydrates formed. It may subsequently be released by respiratory metabolic processes by the producing plant or by other organisms in the food chain. Animals use this source of energy when they perform muscular work. Energy from the sun cascades through plant and animal communities as it passes from one organism to another, supplying life-supporting energy for all.

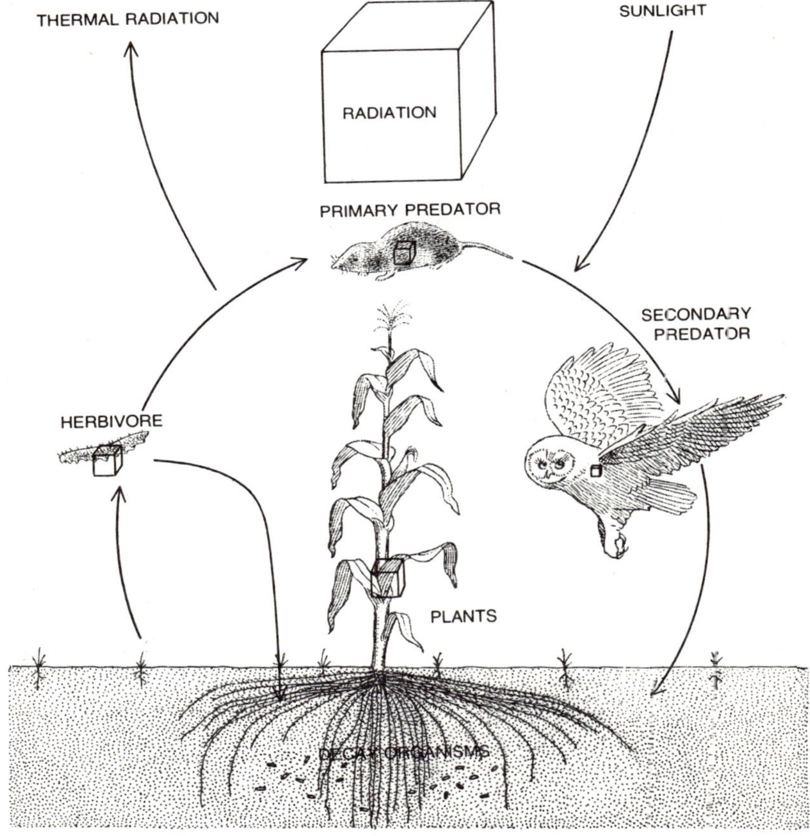

From Janick et al. (1976) with permission of Scientific American

FIG. 2.2. ENERGY CYCLE OF THE BIOSPHERE

Only a very small portion of the total solar energy incident upon a plant leaf is converted to chemical energy by photosynthesis. One estimate of the maximum potential net (photosynthesis minus respiration) plant production is 1.11 $MJ \cdot m^{-2} \cdot d^{-1}$ (71 $g \cdot m^{-2} \cdot d^{-1}$) or 5.3% of the total incident solar radiation (Gates 1971). This is 12% of the visible portion of the solar spectrum. Actual production and conversion rates are considerably less, up to about 1% of total incident solar radiation. Gorz *et al.* (1978) found that 0.8% of the annual solar radiation was fixed in plants by photosynthesis in a New Hampshire forest; net production was about 0.4%.

Energy stored in plants and in animals consumed by other animals is used by the consumer. Only a small portion usually contributes to an increase in weight of the consumer, typically in the order of 10% (Gates 1971; Odum 1971). Janick *et al.* (1976) gave a range of from 2 to 18%. The remainder supports ongoing body functions, such as maintenance, movement and reproduction.

There are two ways of interpreting the energy flow as it proceeds from plants to higher forms of animal life. One is that the energy is concentrated and collected as it flows to a relatively few higher organisms. One might imagine a funnel pouring the net productivity of many organisms in an ecosystem into these few organisms. An alternative interpretation is that most of the energy is consumed along the way and that only a small portion of the energy, but of higher quality, reaches these organisms.

HUNTING AND GATHERING SOCIETIES

Energy is required to produce food and food contains energy. In order to gain an appreciation for the forms and quantities of energy flows which result in food, we shall examine several examples of food production which man has utilized in the past or now utilizes.

Man requires about 10–15 $MJ \cdot d^{-1}$ of food energy, the amount varying with sex, age and size, physical activity, etc. Passmore and Durnin (1955) presented a comprehensive review of food energy needs. However, due to the inefficiencies of conversion of energy from one form to another, more than 10–15 $MJ \cdot d^{-1}$ is ultimately required to provide food for a person. The ultimate requirement may consist of nonrenewable sources, renewable sources, or a combination of the two.

Prior to the development of agriculture (cultivation of crops and raising of livestock useful to man) food was obtained by hunting, fishing and gathering. Few examples of hunting-gathering societies yet exist, and those which do are likely atypical because they are the few which have been left or pushed to the least desirable geographic areas. Hunting-gathering societies are characterized by a total energy consumption not much greater than the nutritive energy of the food they consume.

Lawton (1973) analyzed and gave guidelines for the energetics of food gathering. For an animal to survive, the energy in the gathered food must be at least equal to the energy expended during its collection. To allow energy for other functions, Lawton wrote that the ratio of the two energies should be at least two. He found, for several species, that the ratio of energy gained to energy expended ranged from slightly greater than one for damselfly larva to 70 for a tropical hummingbird. He further cited a human society on a Pacific atoll having a ratio of 18.4.

Lee and DeVore (1968) edited a book in which descriptions of various hunting-gathering societies are presented. It includes a description by Lee of the !Kung Bushmen of the Kalahari desert of Botswana, who are able to provide their food needs by about 60% of their population spending approximately 12 to 19 hours per week hunting and gathering food; or, in other words, an average of 6% of the total time per 24-hour day is required in getting food. The remainder is used for resting and in providing their housing and clothing requirements, and in leisure and socializing. (For additional information *see* Lee 1969.)

A tribe of modern-day Indians of eastern Nicaragua was studied by Nietschmann (1972). They hunt, fish and farm. He found the energy of their catch per hour of activity averaged 5.32 $MJ \cdot h^{-1}$ for hunting and 4.03 $MJ \cdot h^{-1}$ for turtle fishing.

The Batak, a group of Philippine forest hunter-gatherers, were found to obtain 2.02 and 7.28 $MJ \cdot h^{-1}$ from digging two species of wild yams (Eden 1978). Gathering mollusks yielded 1.32 $MJ \cdot h^{-1}$ and jigging for eels yielded 2.27 $MJ \cdot h^{-1}$.

Kemp (1971) provided a description of a modern Eskimo society. This group is quite interesting given their ability to survive in a hostile environment. However, it has been sharply influenced and is greatly energy subsidized by the outside world (ammunition, fuel, food, snow-mobiles, outboard motors, etc.). Total energy consumption is approximately 78 $MJ \cdot caput^{-1} \cdot d^{-1}$.

Cottrell (1955) briefly described the American plains Indian and suggested that food gathering-hunting societies lived a rather precarious existence.

Clark and Haswell (1970) gave a more detailed description of several such societies, pointing out land requirements and also the beginnings of agriculture.

PRIMITIVE AGRICULTURAL SOCIETIES

For various reasons, man changed from hunting-gathering societies to primitive, primarily agricultural societies in which plants are cultivated and/or animals are raised as the predominant food sources. The main reason for the transition to agriculture may have been population pressures, with undoubtedly other reasons involved.

Primitive agricultural societies include a broad range of primarily subsistence (little or no surplus) agricultural systems from the cut and burn farming of the Tsembagas (Rappaport 1971) to highly advanced civilizations such as the Babylonians and Egyptians whose agriculture was dependent upon irrigation. Included are systems in which there were no animals, those which utilized draft animals for increased power above human abilities, and those in which livestock rather than crops were predominant. All are characterized by their energy flows, which are primarily of solar origin, either directly utilized in photosynthesis of the crops or livestock forages, or indirectly through application of human labor and draft animal power which also is derived from the crops and forages grown on the farm. Little cultural energy or energy subsidy is found in primitive agricultural systems.

Black (1971) developed a parameter for evaluating energy flows in agricultural systems which he used to compare several categories of agriculture and which has since been used extensively by others. His *efficiency ratio* is defined as the ratio of output energy in the food produced to the sum of the input energy in human, animal or other forms. Two characteristics of this ratio should be understood. First, the numerator and denominator, although both expressed in energy units, are often different forms of energy. The numerator is the caloric or food energy of the agricultural product and the denominator may be fossil fuels. Second, the ratio is not a true efficiency because all inputs are not included in the denominator. Black's efficiency ratio is analogous to the coefficient of performance of a heat pump (Fig. 2.3). In each, one energy source causes another energy source to be moved or used and each expresses a ratio of the output to one of the energy sources. Black's energy efficiencies vary from three to 34 for nonanimal power systems, and from six to 14 for total energy efficiency for animal power systems.

Agricultural societies are characterized by population densities greater than those of hunting-gathering societies. Conklin (1969) listed population densities from 10 to 25 caput $\cdot$ km^{-2}, Odum (1971) gave two examples at 270 and 2470 caput $\cdot$ km^{-2}, and Cottrell (1955) said population densities may be as high as 740 to 990 caput $\cdot$ km^{-2} (persons per sq km; population density).

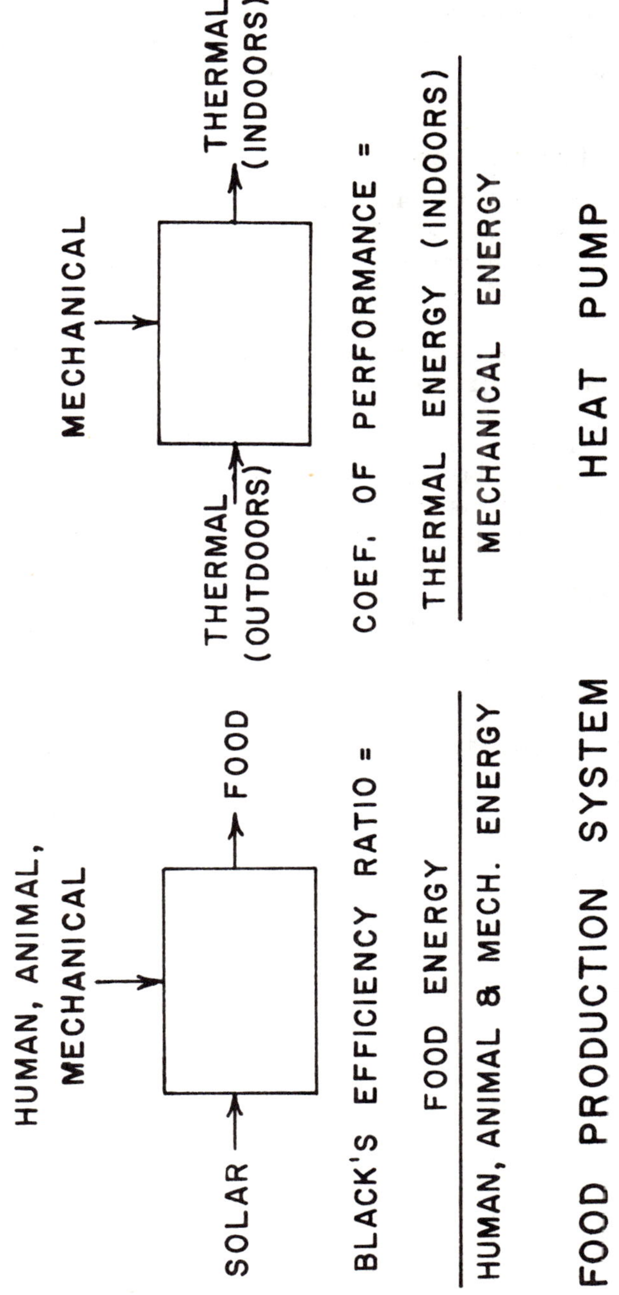

FIG. 2.3. ANALOGY BETWEEN BLACK'S EFFICIENCY RATIO AND COEFFICIENT OF PERFORMANCE FOR A HEAT PUMP

Norman (1978) categorized tropical annual agricultural systems as shifting cultivation, semi-intensive and intensive rainfed, and irrigated. We will subdivide primitive subsistence agricultural systems on the same basis.

Shifting Agricultural Systems

Shifting agricultural systems are characterized by minimal fertilization, lack of livestock and human labor serving as the predominant power input. Cleared land is cultivated for a season or several years and then abandoned to return to forest. When forests are cleared and burned, plant nutrients are added to the soil which would otherwise be added by fertilizers. The energy consumed when a forest is burned may, according to Norman (1978), equal 20 $MJ \cdot kg^{-1}$ of biomass times 100 $t \cdot ha^{-1}$ or 2×10^6 $MJ \cdot ha^{-1}$.

Rappaport (1971) studied a population of primitive New Guinea farmers known as the Tsembagas. He measured their human energy inputs and the agricultural output of their farming operations. Table 2.1 records results for the average of a pair of gardens, one uphill and one downhill, in which different crops are grown due to climatic differences.

The energy efficiency of 17.4 is not atypical for successful examples of primitive agriculture. However, Rappaport did not include the energy of the burned forest when land is cleared for a new garden. If this forest biomass energy is added to the human energy expenditure and is

TABLE 2.1. HUMAN ENERGY EXPENDITURES AND RETURNS FOR GARDENING BY THE TSEMBAGAS

Human Energy Input	$MJ \cdot ha^{-1}$
Clearing underbrush	293
Clearing trees	117
Fencing garden	177
Weeding and burning	98
Placing soil retainers, etc.	75
Planting and weeding until end of harvest	933
Other maintenance	238
Sweet potato harvest	232
Taro harvest	29
Cassava harvest	11
Yam harvest	81
Cartage	619
	2,903
Human food output	31,808
Pig feed output	18,760
	50,568

$$E = \frac{50{,}568}{2{,}903} = 17.4$$

Source: Rappaport (1971).

amortized over the typical two year cropping period before abandonment, the additonal 1×10^6 MJ · ha^{-1} entirely overshadows the human labor energy input of 2903 MJ · ha^{-1}. Instead of an energy efficiency ratio of 17.4, there exists a ratio of 0.05 (for the average of a pair of gardens, $50{,}568 \div (2903 + 1{,}000{,}000)$).

One advantage of shifting cultivation is that human energy inputs are not tied entirely to the planting, cultivating and harvesting of crops with their characteristic peak labor inputs. Clearing may be done during nonpeak labor input periods.

A disadvantage is that croplands are typically scattered and further from the homes, which results in considerable energy expenditure in walking and cartage.

Semi-intensive and Intensive Rainfed Cropping Systems

Semi-intensive and intensive rainfed cropping systems have been studied by many in addition to Black (1971). Leach (1976) has compiled energy budgets from a number of sources for various agricultural systems; energy efficiency ratios are shown for these in Table 2.2.

TABLE 2.2. ENERGY EFFICIENCY RATIOS FOR VARIOUS AGRICULTURAL SYSTEMS

Agricultural System	Energy Ratio
Hunting-gathering	
¡Kung bushmen	7.8
Herders	
Dodo tribe, Uganda	5.0
Shifting cultivators	
Congo	65
Tsembaga	20.3
Subsistence and shifting	
Dayak, rice	16.5–18.2
Iban, rice	14.2
Tanzania, rice	23.4
Maize, Africa	37.7
Millet, Africa	36.2
Sweet potato, Africa	31.3
Cassava, Africa	22.9
Yams, Africa	22.9
Groundnut, Africa	12.8
Subsistence	
India (from Odum)	14.8
China (1935–37)	41.1
Corn, Mexico (axe and hoe)	30.6
(oxen)	14.15
Guatemala (axe and hoe)	13.6
(oxen)	3.95
Nigeria (axe and hoe)	10.5
Philippines (Carabao)	5.07
Wheat, India (bullock)	1.69
Rice, Philippines (Carabao)	5.51

Source: Leach (1976).

Although Leach's figures give no real indication of higher energy efficiency ratios for either shifting or semi-intensive or intensive rainfed (subsistence) agriculture, Norman (1978) suggested human energy input may be somewhat higher for shifting agriculture but that there is no basis for a choice dependent upon energy ratios.

However, there may be some advantage, based on energy considerations, of one crop over another. Chandra *et al.* (1974) computed energy ratios for both subsistence and commercial crops grown by two ethnic groups in a Fiji valley. Ratios ranged from three for tomatoes (commercial) to 130 for bananas (subsistence). The Indian group (Chandra *et al.* 1976) used horses and the Fiji group combined horses and tractors. The Fiji group's crops had a mean energy ratio of 12.8 with a range from two for cauliflower to 23 for Irish potato. The Indian group had a mean energy ratio of 52.2 with a range from three for cucumbers to 130 for bananas. The differences are attributed to the crops grown and the fossil fuel for tractors. The Fiji had a slightly higher economic return of $\$1.07 \cdot h^{-1}$ compared to $\$0.97 \cdot h^{-1}$ for the Indian. This study, therefore, indicates evidence of a conflict between high energy ratio crops and high economic return crops.

Makhijani's (1975) thorough book contains broad background information in the form of six vignettes of Third World agriculture. Mangaon is minimally irrigated, uses draft animals and is basically subsistence. Peipan is highly irrigated, uses draft animals and a minimal amount of mechanical power, and exports a significant portion of its output to Chinese cities. Kilombero is nonirrigated, uses no draft animals and is basically subsistence. Batagawara is nonirrrigated, uses draft animals and is mainly subsistence. Arango is heavily irrigated, uses draft animals and mechanical power, and exports much of its produce to Mexican cities. Quebrada is a single household which uses irrigation, draft animals and sells little of its produce.

Table 2.3's energy efficiency ratio has as the input all forms of energy (human, animal, wood, dung, fossil fuels, etc.) used for agricultural

TABLE 2.3. ENERGY CHARACTERISTICS OF SIX VILLAGES

Location	Size	Total Energy Consumption ($MJ \cdot caput^{-1} \cdot d^{-1}$)	Food Production ($MJ \cdot caput^{-1} \cdot d^{-1}$)	Energy Efficiency Ratio
Mangaon, India	1000	30.8	11.1	0.36
Peipan, China	1000	33.2	23.3	0.70
Kilombero, Tanzania	100	8.7	15.3	1.76
Batagawara, Nigeria	1400	10.5	11.9	1.13
Arango, Mexico	420	130.1	94.6	0.73
Quebrada, Bolivia	6	38.5	7.4	0.19

Source: Makhijani (1975).

purposes. When all energy inputs are included, the energy ratios for those examples of Third World agriculture range from only 0.19 to 1.76.

Also, Makhijani presented several ideas about Third World agricultural systems not emphasized elsewhere:

(1) Total energy usage in Third World agriculture is not small, but rather it is significantly large due to its heavy usage of wood and dung as fuel for cooking and heating.
(2) Energy is used very inefficiently.
(3) Farms in less developed countries may use more energy per unit area than farms in industrialized nations in part because of the energy requirements for feeding draft animals, and they are generally less productive per unit area.
(4) Subsistence farming is often very harmful to the environment; overgrazing and deforestation are worldwide problems but generally most severe in less developed countries.
(5) Third World agricultural systems must have additional energy in the form of useful work, which must be obtained from a combination of more fuel use, fuel substitution and an increase in the efficiency of use of energy resources.

Revelle (1975) also stated a case for increased energy consumption in India's less developed agricultural economy.

Irrigated Primitive Agricultural Systems

Irrigated primitive agricultural systems are generally charcterized by greater human and total energy inputs and outputs than rainfed agriculture. Norman (1978) presented data from several sources which indicate that human labor input is at least doubled in comparison to rainfed, that outputs are also increased sharply, but that energy ratios may be somewhat less with irrigation.

One often significant advantage of irrigation is that in tropical or subtropical areas, multiple cropping is enabled, along with a lesser requirement for food storage and a better distribution of labor inputs.

Energy requirements for activities in the food system subsequent to production are a significant portion of total energy requirements. Makhijani (1975) estimated energy requirements for cooking, in the form of dried cow dung and wood, at 15–20 MJ $\cdot$ caput $\cdot$ d^{-1} for Third World economies. Revelle's estimate for India was 13.5 MJ $\cdot$ $caput^{-1} \cdot d^{-1}$. According to Revelle (1975), efficiency of energy use in cooking was less than 9%, contributing to high rates of energy consumption and deforestation.

Cartage or hauling of food from the point of production to consumption is a significant consumer of human and draft animal energy. Rappaport calculated human cartage energy requirements for the Tsembaga's to carry produce for themselves and their pigs to their homes at $0.15\ \mathrm{MJ} \cdot \mathrm{caput}^{-1} \cdot \mathrm{d}^{-1}$.

BIBLIOGRAPHY

BLACK, J.N. 1971. Energy relations in crop production—a preliminary survey. Ann. Appl. Biol. *67,* 272–278.

CHANDRA, S., DE BOER, A.J. and EVENSON, J.P. 1974. Economics and energetics: Sigatoka Valley, Fiji. World Crops *26* (1) 34–37.

CHANDRA, S., EVENSON, J.P. and DE BOER, A.J. 1976. Incorporating energetic measures in an analysis of crop production practices in Sigatoka Valley, Fiji. Agric. Syst. *1* (4) 301–311.

CLARK, C. and HASWELL, M. 1970. The Economics of Subsistence Agriculture. Macmillan, New York.

CONKLIN, H.C. 1957. Hanunoo agriculture. FAO Forestry Development Ppr. No. *12.* Food and Agric. Organ.—U.N., Rome.

CONKLIN, H.C. 1969. An ethnoecological approach to shifting agriculture. *In* Environment and Cultural Behavior. A.P. Vayda (Editor). Natural History Press, New York.

COTTRELL, F. 1955. Energy and Society. McGraw Hill, New York.

EDEN, J.F. 1978. The caloric returns to food collecting: Disruption and change among the Batak of the Philippine tropical forest. Human Ecol. *6* (1) 55–69.

GATES, D.M. 1971. The flow of energy in the biosphere. Sci. Amer. *24* (3) 48–92, 94, 96–100.

GORZ, J.R., HOLMES, R.T., LIKENS, G.E. and BOUMANN, F.H. 1978. The flow of energy in a forest ecosystem. Sci. Amer. *238* (3) 93–102.

HUBBERT, M.K. 1971. The energy resources of the earth. Sci. Amer. *224* (3) 60–70.

JANICK, J., NOLLER, C.H. and RHYKERD, C.L. 1976. The cycles of plant and animal nutrition. Sci. Amer. *235* (3) 76–86.

KEMP, W.B. 1971. The flow of energy in a hunting society. Sci. Amer. *224* (3) 104–115.

LAWTON, J.H. 1973. The energy cost of "food-gathering." *In* Resources and Population. Bernard Benjamin (Editor). Academic Press, New York.

LEACH, G. 1976. Energy and Food Production. IPC Science and Technology Press, Guildford, Surrey, England.

LEE, R.B. 1969. !Kung Bushmen subsistence: An input-output analysis. *In* Environment and Cultural Behavior. A.P. Vayda (Editor). Natural History Press, New York.

LEE, R.B. and DEVORE, I. 1968. Man, The Hunter. Aldine Publishing Company, Chicago.

LOOMIS, R.S. 1976. Agricultural systems. Sci. Amer. *235* (3) 99–105.

MAKHIJANI, A. 1975. Energy and Agriculture in the Third World. Ballinger Publishing Company, Cambridge, Mass.

NIETSCHMANN, B. 1972. Hunting and fishing among the Miskito Indians, Eastern Nicaragua. Human Ecol. *1* (1) 41–67.

NORMAN, M.J.T. 1979. Annual Cropping Systems in the Tropics, An Introduction. University of Florida Press, Gainesville. (in press)

ODUM, H.T. 1971. Environment, Power, and Society. John Wiley and Sons, New York.

ODUM, H.T. and ODUM, E.C. 1976. Energy Basis for Man and Nature. McGraw-Hill, New York.

PASSMORE, R. and DURNIN, J.V.G.A. 1955. Human energy expenditures. Physiol. Rev. *35,* 801–840.

RAPPAPORT, R.A. 1971. The flow of energy in an agricultural society. Sci. Amer. *224* (3) 116–122, 127–132.

REVELLE, R. 1975. Energy use in rural India. Science *192,* 969–975.

WATT, B.K. and MERRILL, A.L. 1963. Composition of Foods. Agric. Hdbk. No. *8,* U.S. Dep. of Agric., Washington, D.C.

3

Energy Use in Industrialized Food Production

Modern industrialized systems such as in the United States attracted special attention after the energy shortages in the early 1970s. Realization that food supplies are directly dependent on energy supplies became widespread. At the same time, agriculture was criticized for being particularly inefficient in terms of fossil fuel energy inputs per unit of food energy produced. Since then, many proposals for reducing on-farm energy inputs, such as eliminating the use of manufactured fertilizers and returning to animal power, have been given. Before such decisions can be made there should be a thorough analysis of all energy flows in the food system.

Americans have recently begun to reexamine those essential things which constitute the basic fabric of life such as ample food, shelter and energy supplies. These are necessities whose availability most Americans have grown to accept without question. Americans have been fortunate because this nation has not known hunger on any massive scale in modern times, but this is not true of much of the rest of the world.

Malthus, in his 1798 classic essay on population and the future of mankind, predicted that man's food needs would outstrip his ability to produce food. He properly forecast the population growth, but he did not see the large increases in agricultural productivity made possible through scientific technology. Fortunately, agricultural scientists interceded to show the way for agricultural production to increase at rates not imagined even by brilliant men such as Malthus.

New warnings are being sounded about the future. Generalized computer models warn of the limits of growth and of future catastrophe.

These warnings should not be discounted, but rather one should carefully examine the assumptions which are a part of these analyses.

In the search for a solution to the world's food problems, the common attempt to transplant a small piece of a highly industrialized food system to the hungry nations of the world is plausible, but the long-term benefits are thus far unclear. An examination of the energy flows in the United States food system as they have developed may provide some insights to the problem.

A trend in United States agriculture has been toward the concentration of farm production on fewer but larger farms worked by a declining number of farmers. Nevertheless, farm production has increased, largely due to technological improvements which have enabled modern farmers to substitute capital and other inputs for labor.

United States farmers have become very efficient food producers. Using fossil fuel inputs, Stickler *et al.* (1975) reports that modern agriculture can achieve solar energy capture efficiencies 20 times greater than primitive agricultural systems. Even with this increased efficiency, only a small part of the incident solar energy is eventually converted to food energy and consumed by humans. For example, 28,000 units of solar energy reaching the outer limits of the atmosphere in the United States result in only one unit of food energy on the table. The atmosphere filters out all but 16,000 units and less than half of this is in the photosynthetically active range. Much of this misses the plants or is reflected. Even the portion absorbed is converted only partially to chemical energy in the form of food, feed or fiber. Figure 3.1 shows the distribution of the 16 units of plant energy resulting from the 28,000 units of solar energy reaching the outer limits of the atmosphere. Considering only that energy reaching the earth's surface in the photosynthetically active range, the 16 units of plant energy represent a 0.2% efficiency, which is the average for the United States crops and grazing land. As noted previously many crops have an efficiency of more than 1%.

Stickler *et al.* (1975) reported that large farms utilizing large tractors and farm machines are more efficient than smaller ones. Farms smaller than 40 ha use 90% more fuel per hectare for field operations than do farms larger than 100 ha (Stickler *et al.* 1975).

Modern farmers use chemical fertilizers, herbicides and insecticides to help them capture solar energy in plants efficiently. Some 90% of the energy used in nitrogen fertilizer manufacture comes from natural gas. The amount of natural gas so used represents only 2% of the United States energy budget. Yet ⅓ of the total United States food production can be attributed to fertilizers.

Prior to 1920 and before the tractor era, more than 26 million draft animals powered United States agriculture. To provide their feed re-

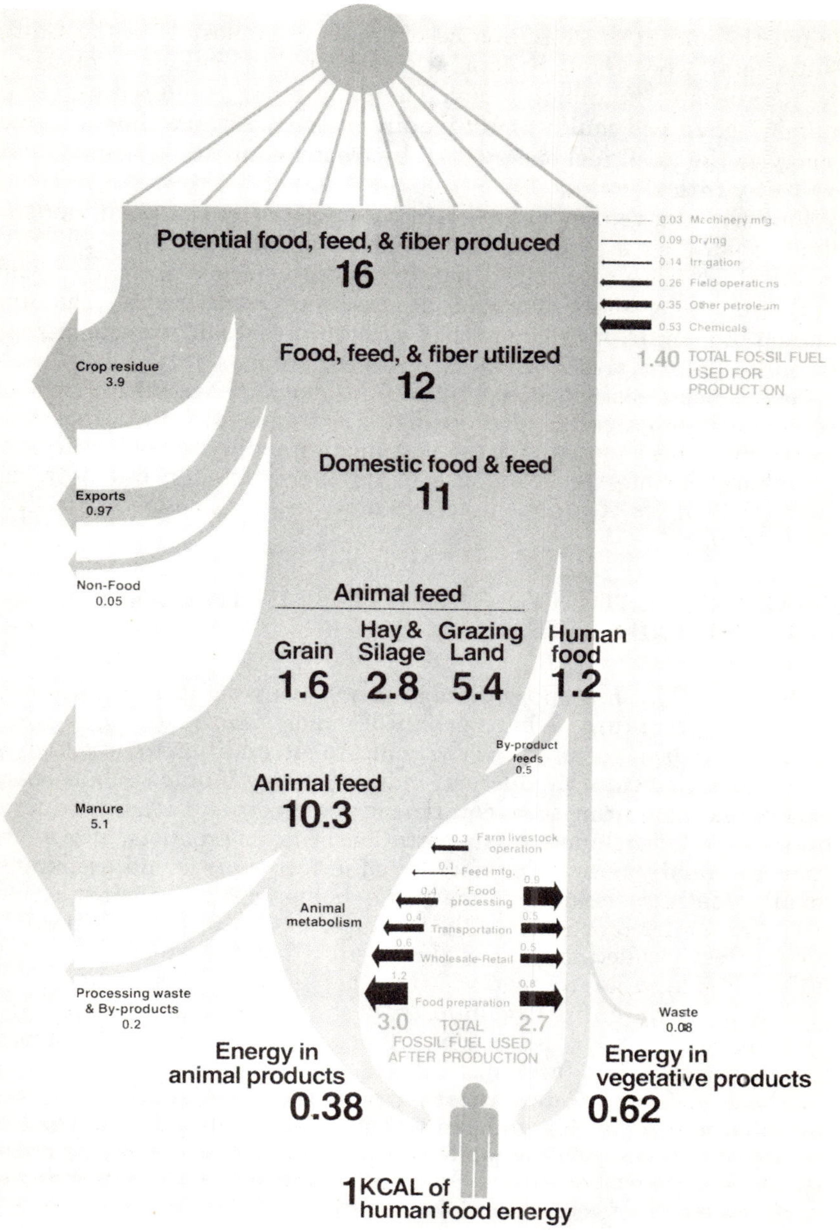

From Stickler et al. (1975)

FIG. 3.1. ENERGY REQUIRED TO PRODUCE ONE UNIT OF HUMAN FOOD ENERGY

quired the output of 40 million ha of cropland, almost ⅓ of the United States total, when the U.S. population was only 100 million compared to the more than 200 million today. The plant energy required to feed 26 million horses and mules would require more than three times as much energy as the fossil fuel energy used by tractors and other farm machinery today (Stickler *et al.* 1975).

Similarly, a given unit of manure contains far less nutrient value for plants than does a given unit of chemical fertilizer. It takes 207 kg of chemical fertilizer to supply 1 ha of corn with sufficient nitrogen while 1.17×10^5 kg of manure would be required. Furthermore, the total amount of manure produced in the United States would provide nitrogen for only ⅓ of the present United States corn crop.

Thus, some apparent alternatives to mechanized agriculture not only would be impractical if adopted in the extreme, but are also energy-inefficient. This is not to suggest that modern industrialized agriculture is making the most efficient use of its energy inputs, but only that thorough analysis of various practices must be made before alternatives can be adopted.

CHARACTERISTICS OF ENERGY USE IN THE UNITED STATES FOOD SYSTEM

First of all, it seems appropriate to define just what is meant when one speaks of agriculture. What portion of energy use in the agribusiness chain should be associated with agriculture? In addition to direct energy inputs in agricultural production, indirect energy inputs such as energy used in manufacturing farm equipment, fertilizers and other production inputs may be included. Transportation of farm products, processing, trade and finally household energy used in food storage and preparation are all sometimes included in the energy budget charged against agriculture. This entire process from the producer to the consumer is probably more correctly called the *food chain.* Figure 3.2 depicts some components of this chain and its interaction with other sectors of the economy. The food chain, or the entire food industry of the United States, is intricately interwoven with many sectors of the economy and obviously involves much more than just farming.

A Federal Energy Administration publication (FEA 1976) by Booz, Hamilton and Allen, Inc. indicated that approximately 16% of the total United States energy requirement is used in the food system. Earlier estimates by Hirst (1973) and Steinhart and Steinhart (1974) indicated lower values of between 12 and 14%. Table 3.1 gives the breakdown for each component of the food chain.

In order to make a realistic accounting of energy associated with any sector of our economy it is important that all (direct and indirect) energy

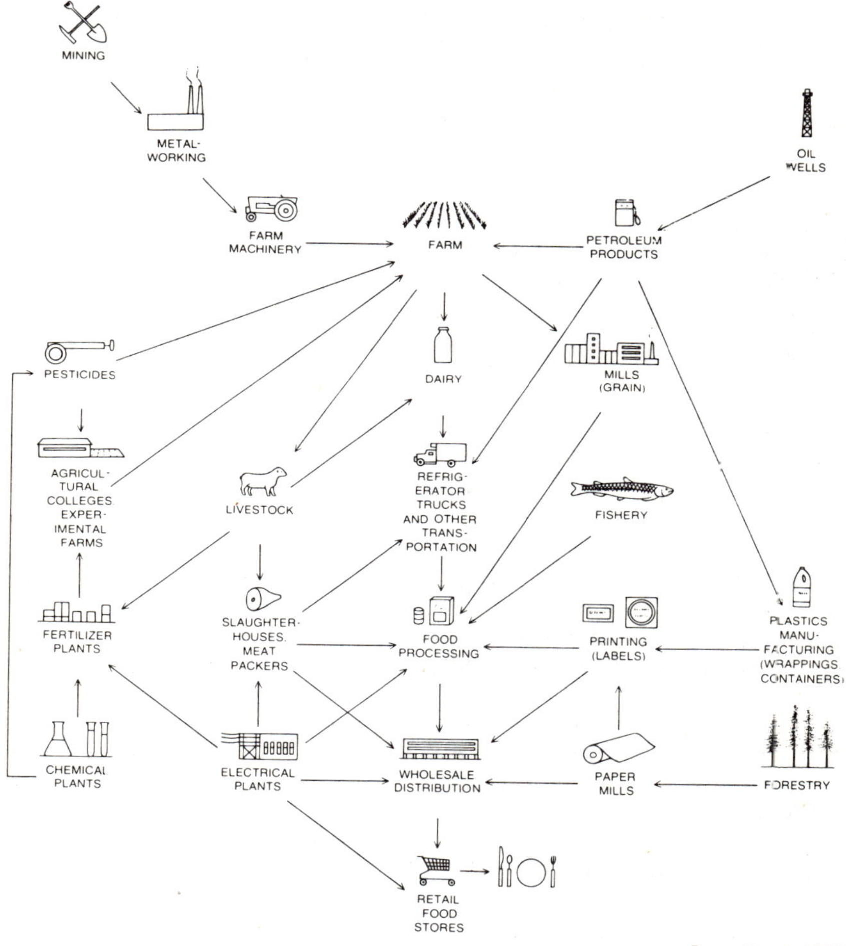

From Heady (1976)

FIG. 3.2. FOOD CHAIN AND ITS INTERACTION WITH OTHER SECTORS OF THE ECONOMY

inputs be included. For example, in production agriculture this includes indirect energy use such as that required to produce farm machinery, including energy to manufacture steel for the machines, and the energy used to manufacture fertilizer, pesticides, plastics and other production inputs. Yet with all these inputs included, production agriculture makes up less than 3% of the total United States energy consumption.

TABLE 3.1. FOOD SYSTEM ENERGY CONSUMPTION AS PERCENTAGES OF TOTAL UNITED STATES ENERGY CONSUMPTION[1]

Component	Direct	Indirect	Capital	Transportation	Total
Production	1.0	1.1	0.4	0.4	2.9
Manufacturing	1.8	2.5	0.1	0.4	4.8
Distribution					
Wholesale trade	0.5		—	—	0.5
Retail trade	0.8		—	—	0.8
Consumption					
Out-of-home preparation	2.0	0.2	—	0.6	2.8
In-home preparation	3.3	—	0.3	0.7	4.3
Totals (Food system)	13.2		0.8	2.1[2]	16.1

Source: FEA (1976)
[1]Compiled from data covering differing years from 1963 to 1971.
[2]Excludes energy use for transportation in retail and wholesale trade which is included in the "direct" and "indirect" estimates for these sectors.

Although the amount of energy consumed in the United States food chain might be termed relatively low, compared to the total United States consumption, it has been increasing at a rapid rate as indicated in Fig. 3.3. Note that the energy content of the food consumed increased at a much slower rate.

Figure 3.4 displays features of our food system not easily seen from economic data. The curve gives an account of our history of increasing food production and, like most growth curves, suggests that additional energy input to the food system is not likely to yield significant increases in production. A similar relationship exists between additional energy input and reduction of labor input.

Table 3.2 shows a number of per capita food indicators. Per capita expenditures increased significantly, but much of this was due to

TABLE 3.2. PER CAPITA FOOD TRENDS IN THE UNITED STATES

	Expenditures ($ · yr^{-1})	Consumption (kg · yr^{-1})	Food energy (MJ · d^{-1})	Protein (g · d^{-1})
1960	448	653	13.1	95
1963	468	644	13.2	96
1965	510	644	13.1	96
1967	549	649	13.4	98
1970	648	658	13.8	100

Source: Hirst (1973).

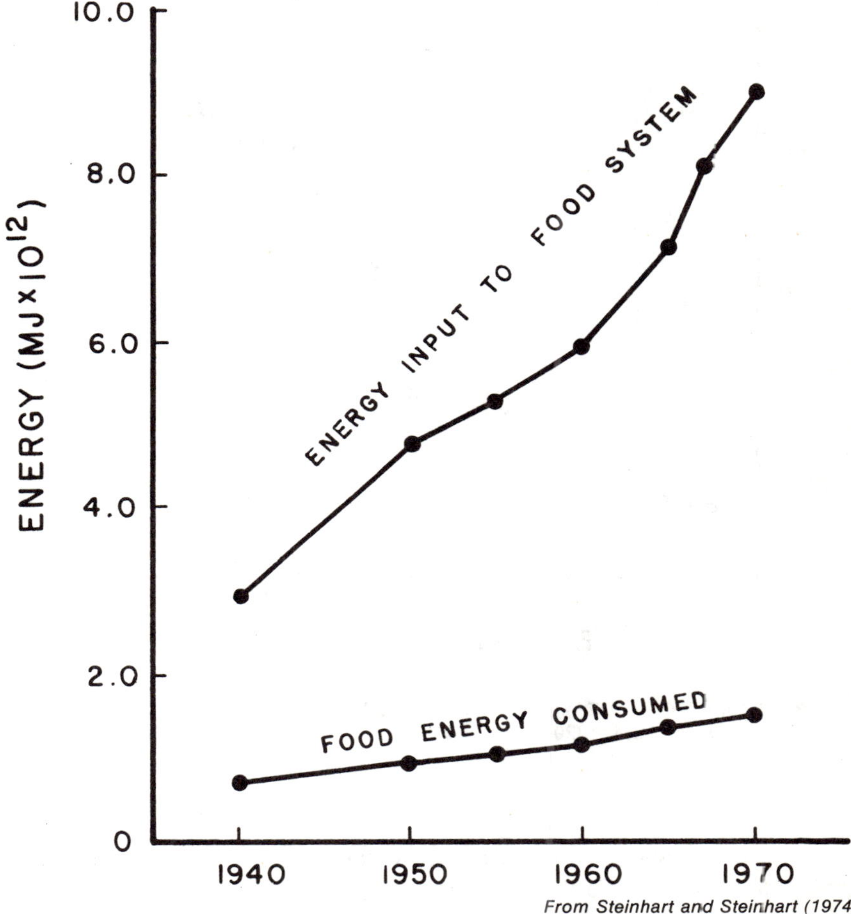

From Steinhart and Steinhart (1974)

FIG. 3.3 ENERGY USE IN THE FOOD SYSTEM, 1940–1970, COMPARED TO THE CALORIC CONTENT OF FOOD CONSUMED

inflation. Deflating the expenditure figures by the food price index shows a 25% increase in real food expenditures. This increase is primarily due to a shift towards more expensive foods such as beef, processed foods and consumption of food away from home.

The United States food consumption is not only shifting towards more expensive foods, but also towards more energy intensive foods. Figure 3.5 shows the ratios of primary energy use to food energy content for major

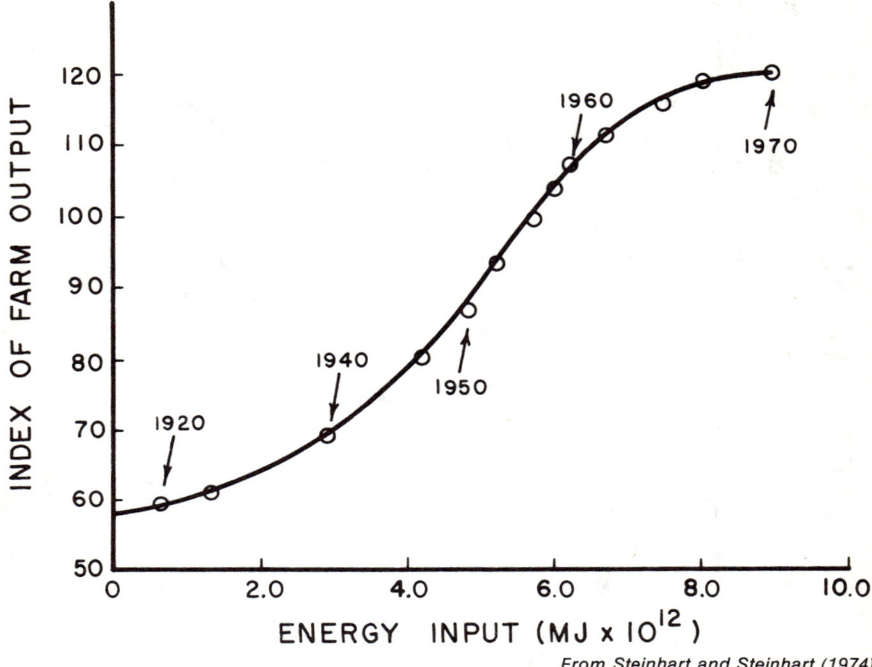

From Steinhart and Steinhart (1974)

FIG. 3.4. FARM OUTPUT AS A FUNCTION OF ENERGY INPUT TO THE UNITED STATES FOOD SYSTEM, 1920–1970

food groups. However, comparing the food energy content of various food groups may be of little value, particularly for United States consumers, since other characteristics of food appear to be more important, such as protein, vitamin and mineral contents, palatability, and maybe even the lack of calories. Figure 3.6 gives the distribution of primary energy consumption for food by major food groups.

Pimentel (1973) reported that the average annual per capita expenditure for food in the United States was about $600 in 1970—a figure which is very high by world standards, but still represents less than 20% of our disposable income. It is interesting to compare the percentage of income spent on food with the percentage of the total energy budget associated with the food system. Presently the energy value is slightly less than the dollar value. How will further increases in the cost of energy affect these values?

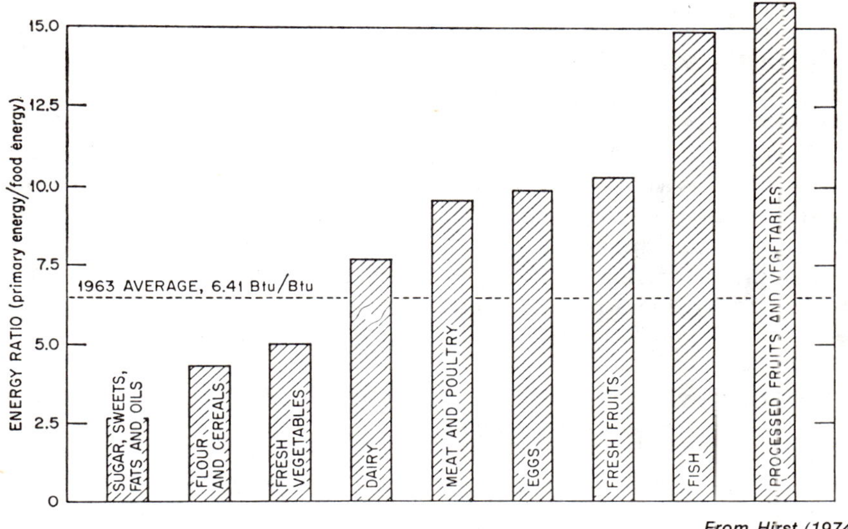

From Hirst (1974)

FIG. 3.5. RATIO OF PRIMARY ENERGY USE TO FOOD ENERGY CONTENT FOR MAJOR FOOD GROUPS, 1963

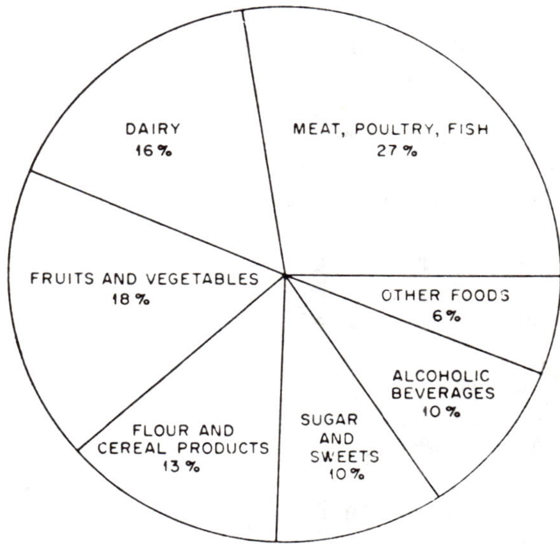

From Hirst (1974)

FIG. 3.6. DISTRIBUTION OF PRIMARY ENERGY USE IN THE UNITED STATES FOR FOOD BY MAJOR FOOD GROUPS, 1963

Individual Economic Sectors

Production Agriculture.—Most sources show that production agriculture (primarily on-farm operations) accounts for less than 20% of the total energy required in the food system. Home preparation consumes more energy than production agriculture or food processing (Table 3.1). Figure 3.7 gives a comparison of energy inputs into crop production in the United States. One way of identifying and understanding the importance of energy inputs into crop production is to consider a specific crop that is fairly representative of United States agriculture, such as corn. Pimentel *et al.* (1973) made a relatively thorough analysis of United States corn production. Table 3.3 lists the various inputs to corn production in terms of MJ · ha^{-1}.

While corn yields increased about 240% from 1945 to 1970, the labor input per hectare decreased more than 60%. Machinery in agriculture

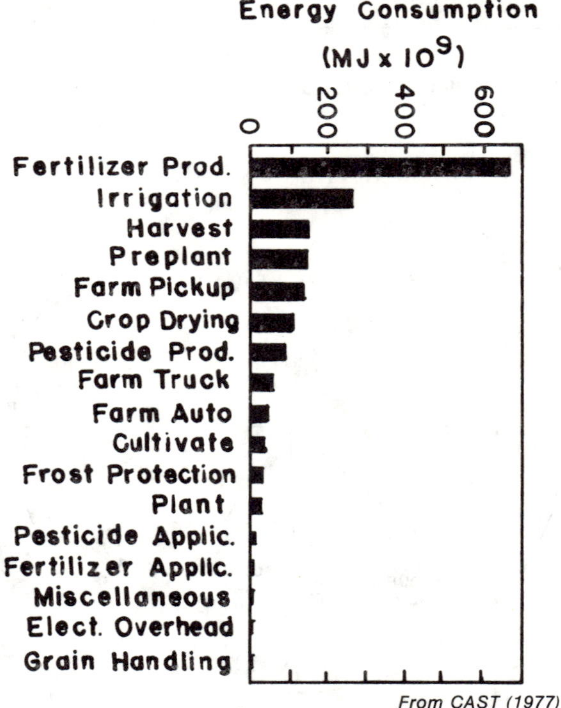

From CAST (1977)

FIG. 3.7. CONSUMPTION OF ENERGY IN DIFFERENT ASPECTS OF CROP PRODUCTION IN THE UNITED STATES, 1974

has increased significantly during the past 20 years; the mean rate of power per farm worker has increased from 7.5 to 35 kW. For total United States corn production, fuel consumption for all machinery rose from 140 L ha^{-1} in 1945 to about 206 L ha^{-1} in 1970. Also, the use of fertilizer has been rising steadily since 1945. The increase in nitrogen has been about 16-fold.

Hybrid corn that is currently harvested has a higher moisture content because the newer varieties have growing seasons which extend further into the fall when drying conditions are poor. This factor along with increased awareness of the benefits of drying have contributed to the tenfold increase in drying energy over the period from 1945 to 1970 (Table 3.3). Energy for irrigation represents a relatively small input for corn on the average since only about 4% of the corn was irrigated in 1964. Thus, for irrigated corn the energy input per hectare would be about 10,000 MJ, representing the largest single input. Some crops located in dry climates require many times this amount of energy for irrigation.

The 30,000 MJ $\cdot$ ha^{-1} input of fossil fuel represents a small portion of the energy input when compared with the solar energy input. During the growing season, about 21×10^6 MJ reach a 1 ha cornfield; about 1.3% of this is converted into corn and about 0.4% into corn grain. The 1.3% represents about 0.275×10^6 MJ, making the fossil fuel input about 11% of the total.

Food Processing.—Many farm products are not in a form desirable for human consumption and must be processed before they are sold in the retail market. Other products are highly perishable and must be preserved in some manner if they are to be available for future consumption. Processing of agricultural products for human consumption in an industrialized food system requires a considerable amount of energy. As noted previously, more than 80% of the energy required in the United States food system occurs after the product leaves the farm and a large portion of this goes into processing. According to the Economic Research Service (ERS 1974), food processing or manufacturing required about 8% of the energy used in all manufacturing industries.

The amount of energy consumed in food processing has steadily increased—an increase of 49% from 1954 to 1972. This is partly due to a general increase in energy used in all manufacturing, but is due largely to a shift toward more energy intensive processing of food before it is retailed. This may not in every case result in an increase in energy for the complete food chain because less energy may be required in home preparation due to the additional processing.

TABLE 3.3. ENERGY INPUTS ($MJ \cdot ha^{-1}$) IN CORN PRODUCTION[1]

Input	1945	1950	1954	1959	1964	1970
Labor[2]	129	101	96	79	62	51
Machinery	1,862	2,586	3,103	3,620	4,344	4,344
Gasoline	5,621	6,370	7,120	7,494	7,868	8,244
Nitrogen	608	1,303	2,346	3,562	5,039	9,731
Phosphorus	110	157	188	251	283	487
Potassium	54	109	521	624	703	703
Seeds for planting	352	418	195	378	314	652
Irrigation	197	238	279	321	352	352
Insecticides	0	11	34	80	114	114
Herbicides	0	6	11	29	43	114
Drying	103	310	621	1,034	1,241	1,241
Electricity	331	559	1,034	1,448	2,100	3,207
Transportation	207	310	465	621	724	724
Total inputs	9,574	12,478	16,013	19,241	23,187	29,964
Corn yield (output)	35,450	39,620	42,750	56,300	70,900	84,450
Output/input	3.70	3.18	2.67	2.88	3.06	2.82

Source: Pimentel *et al.* (1973).

[1]Authors give a very complete discussion and source of data for this table.

[2]In this analysis, energy for labor is taken as a portion of the food energy consumed by the worker. As will be pointed out in Chapter 4, this figure does not represent the total amount of energy associated with labor.

Table 3.4 gives the energy consumption by the major food processing industries and the value of the product. Approximately half of the energy required in food processing is used in the production of inputs consumed in the manufacturing process, such as metal for cans, and for production of capital inputs. The amount of energy associated with these inputs has probably been understated since few data are available on capital inputs.

A study by Singh (1978) stresses the importance of more precise energy accounting in the food processing industries in order to identify potential areas for reducing energy consumption. He pointed out that excessive energy currently is being expended to insure compliance with some food regulations.

Brown and Batty (1976) analyzed the energy required to process a can of corn. Husking, removing from cob, washing and placing in can, and cooking required 2.74 MJ $\cdot$ can^{-1}. The energy required in manufacturing the steel can was 4.17 MJ $\cdot$ can^{-1} and the carton in which the cans were placed amounted to another 0.45 MJ $\cdot$ can^{-1}. Altogether, the processing and packaging amounted to about four times as much energy as that

TABLE 3.4. ENERGY USE AND SALES ASSOCIATED WITH FOOD PROCESSING

Industry Name	Sales (billion $)	Energy Use (billion MJ)
Meat products	13.7	823
Creamery butter	0.7	54
Cheese, natural and processed	0.7	47
Condensed and evaporated milk	0.4	38
Ice cream and frozen desserts	1.2	73
Fluid milk	5.2	330
Canned and cured sea foods	0.3	15
Canned specialties	0.9	70
Canned fruits and vegetables	2.3	178
Dehydrated food products	0.2	13
Pickles, sauces and salad dressings	0.6	42
Fresh or frozen packaged fish	0.3	15
Frozen fruits and vegetables	1.4	99
Flour and cereal preparations	1.2	83
Prepared feeds for animals and fowl	0.5	39
Rice milling	0.1	8
Wet corn milling	0.02	2
Bakery products	5.6	294
Sugar	0.9	71
Confectionery and related products	1.8	104
Alcoholic beverages	5.5	251
Bottles and canned soft drinks	2.0	118
Flavoring extracts and sirups	0.2	13
Roasted coffee	1.5	38
Shortening and cooking oils	0.7	66
Macaroni and spaghetti	0.2	12
Other miscellaneous foods	1.7	108
Totals	51.5	3004

Source: Hirst (1973).

required to produce the corn. The manufacturing of the can required 1.4 times as much energy as the corn production.

Proving that packaging food requires a far greater use of energy resources than does the growing of that food does not automatically lead to practical alternatives for preservation. Freezing rather than canning food is one option that warrants analysis, but many factors must be considered such as length of time the product is held before consumption. For example, the energy required to keep a 1 kg package of frozen corn is about 0.188 $MJ \cdot d^{-1}$. Accounting only for the electricity to operate the freezer, corn maintained longer than 22 days requires more energy than canned corn; holding for one year requires 17 times as much energy.

Transportation, Marketing and Household Preparation.—About half of the energy consumed in the food system goes for transportation, marketing and household preparation.

Spacious, comfortable supermarkets, open 24 hours a day with long rows of open refrigerated display cases are designed to tempt and entertain the buyer rather than efficiently distribute food.

Incredible as it may seem, the data indicate that consumers spend more energy getting their food from the supermarket to their kitchens than the farmer uses to produce it. Consumers also use more energy in home preparation than was used on the farm. It is also interesting to note that people will burn 800 MJ in the form of gasoline in order to consume an 8 MJ meal in a restaurant.

To move fresh and processed foods through the various marketing channels, certain essentials are required. The uninterrupted flow of farm products is vital to the health and welfare of everyone. Without adequate transportation, marketing and storage, much food will be lost through spoilage. Spoiled food near the end of the food chain represents a tremendous energy loss since most of the energy inputs have been made.

Proposed changes in the food system must be carefully analyzed. For example, it would be futile to ship fresh produce by boat instead of truck if it spoiled before reaching its destination, even when trucking requires five times as much energy. Table 3.5 gives the energy required for transporting food in the United States.

To summarize the energy inputs in the food chain, consider the inputs to a can of corn (Fig. 3.8) as reported by Brown and Batty (1976).

Types of Fuel Used in the United States Food System.—Fifty percent of the energy used in the United States food system in 1970 was petroleum—primarily diesel fuel, gasoline and liquefied petroleum (LP) gas. Natural gas supplied 30% and electricity 14%. Table 3.6 illustrates further distribution of energy consumed in the United States food

TABLE 3.5. PROPULSION ENERGY REQUIREMENTS FOR FOOD TRANSPORT, 1963

Mode	Amount Shipped (Mt)	Amount Shipped (Gt · km)	Average Haul Length (km)	Propulsion Energy Use (PJ)	Energy Intensiveness ($MJ \cdot t^{-1} \cdot km^{-1}$)
Railroads	85.0	73.9	869	40.09	0.54
Trucks	98.0	35.9	370	63.83	1.78
Boats	2.4	2.0	837	0.74	0.37
Totals	185.4	111.8	595	104.66	0.94

Source: Hirst (1973).

system, including a comparison between production agriculture and the total for the food system.

The food and fiber sector is facing current shortages of LP gas and natural gas. Processing firms are dependent upon natural gas, and conversion of these plants to coal would be costly and would increase food prices for the consumer. Also, natural gas is very critical to at least three inputs in production agriculture—fertilizers, irrigation and crop drying. Natural gas is used as a feedstock for almost all of the nitrogen fertilizer produced in the United States. Some have suggested that natural gas is so important to fertilizer production that it should not be burned at all but saved for chemical use.

The types of fuel used for various inputs and operations in the United States food system vary considerably with location. For example, irrigation in the southwest depends heavily on natural gas where 40 to 60% of the power units use natural gas. On the other hand, Florida is not so dependent upon natural gas, but requires about 40% of its energy for production agriculture in the form of fuel oil primarily for crop heating.

TABLE 3.6. FUELS AND QUANTITIES USED IN THE UNITED STATES FOOD SYSTEM IN 1970

Fuel	Production TJ	Production %	Food and Fiber Sector[1] TJ	Food and Fiber Sector[1] %
Gasoline	530,837	47.9	809,910	16.4
Diesel	394,500	35.6	194,003	24.3
Distillate fuel oil			196,640	4.0
LP gas	130,978	11.8	262,344	5.3
Residual fuel oil			102,890	2.1
Natural gas			1,492,253	30.3
Electric power	52,914	4.8	678,403	13.8
Coal			174,892	3.6
Other			12,196	0.2
Total	1,109,229		4,923,531	

Source: ERS (1974).

[1]Includes production, farm family living, processing, marketing and distribution, and manufacturing of inputs

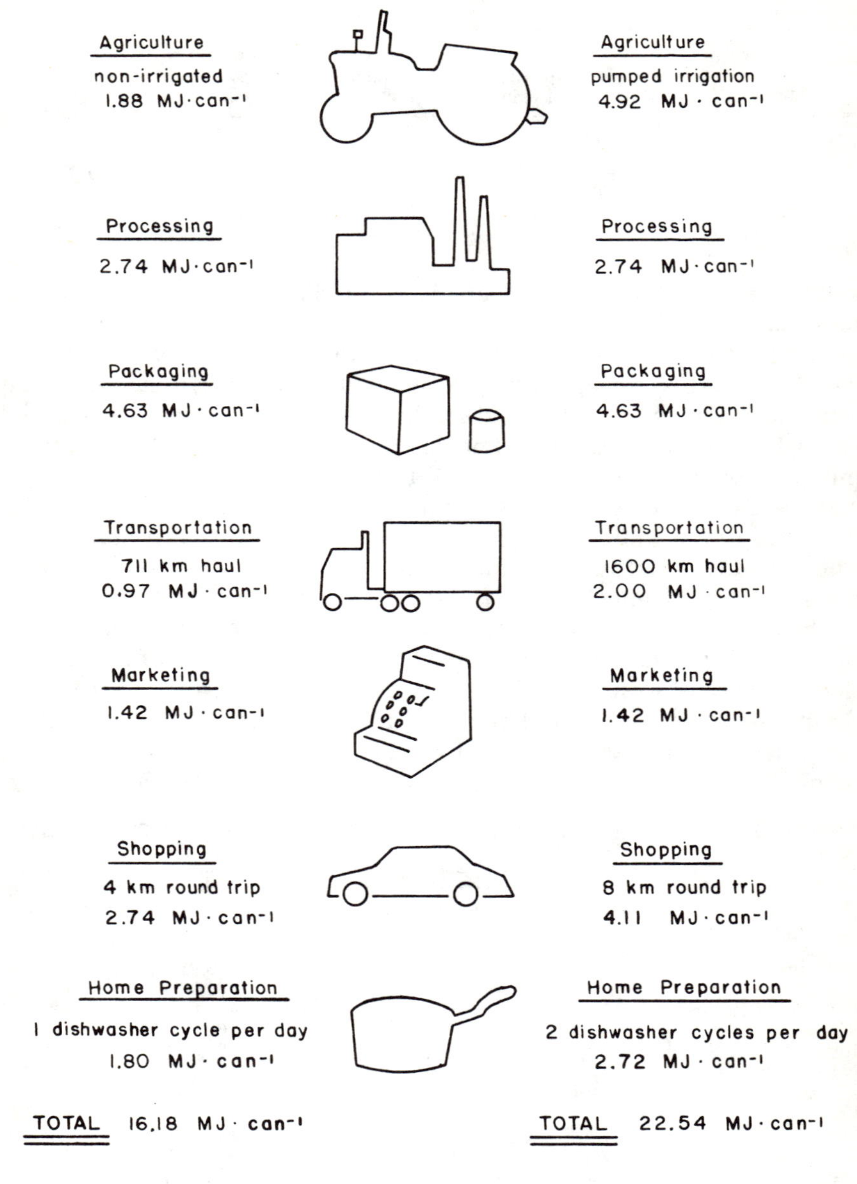

From Brown and Batty (1976)

FIG. 3.8. SUMMARY OF ENERGY INPUTS TO A 0.454 KG CAN OF CORN

TABLE 3.7. PERCENTAGE OF TOTAL ENERGY USE BY FUEL TYPE FOR 14 LEADING ENERGY-USING FOOD AND RELATED PRODUCTS INDUSTRIES FOR 1973

Industry	Natural gas	Purchased electricity	Petroleum products	Coal	Other
Meat packing	46	31	14	9	0
Prepared animal feeds	52	38	10	< 1	0
Wet corn milling	43	14	7	36	0
Fluid milk	33	47	17	3	0
Beet sugar processing	65	1	5	25	4
Malt beverages	38	37	18	7	0
Bread and related products	34	28	38	0	0
Frozen fruits and vegetables	41	50	5	4	0
Soybean oil mills	47	28	9	16	0
Canned fruits and vegetables	66	16	15	3	0
Cane sugar refining	66	1	33	0	0
Sausage and other meat	46	38	15	1	0
Animal and marine fats and oils	65	17	17	1	0
Manufactured ice	12	85	3	0	0

Source: Unger (1975).

Table 3.7 shows data on energy use by fuel type for 14 leading food and related product industries for 1973. Here again, the heavy dependence by agriculture on natural gas is indicated.

Seasonal variations in the demand for certain types of fuel may also complicate any problems that may arise from fuel shortages or inadequate distribution. Many of the production operations require a large portion of their annual energy budget or fuel demand over a short period of time, such as for planting, irrigation, harvesting, drying and freeze protection. Very little data on seasonal variation by fuel type are available. A report from Florida (Smerdon *et al.* 1975) gives "three month moving averages" for each month for gasoline, LP gas and fuel oil. Gasoline requirements in agricultural production remained relatively constant throughout the year, while LP gas requirements in June (during tobacco curing) were over seven times that required in October. Fuel oil requirements were double the yearly average during the heating season. Year to year variation in heating requirements for freeze and frost protection may cause large variations in the annual requirements for production agriculture. A U.S. Department of Agriculture—Federal Energy Administration (1977) publication indicates that 60.8% of the energy required in United States citrus groves was for frost protection, but yet heat for frost protection may not be required at all for several years. Thus, any attempt to allocate fuel based on average values could be disastrous to agriculture.

ENERGY USE IN OTHER INDUSTRIALIZED FOOD SYSTEMS

The characteristics of energy use in industrialized food systems other than the United States appears to be similar to that of the U.S. in that fossil energy has been used to increase production and capital has been substituted for labor. The main differences appear to be the degree to which these practices have been adopted in various countries.

Energy use in food systems has been presented in several ways—energy per unit of food produced, energy per capita, total energy used, etc. The latter two can be misleading when used to compare the effectiveness of energy inputs in various countries. For example, in comparing energy use in the United States and the United Kingdom crop production, Green (1978) gives figures of 7.6 $GJ \cdot yr^{-1}$ per person in the U.S. and 4.3 $GJ \cdot yr^{-1}$ per person for the U.K. At first, it appears that energy input for crop production to feed one person is greater in the United States. However, considering that the U.S. produces enough food for 370 million people (almost twice the population) and the United Kingdom produces enough food for 25 million people (less than half their population) results in adjusted figures of 4.5 $GJ \cdot yr^{-1}$ per person for the U.S. and 9.4 $GJ \cdot yr^{-1}$ per person for the U.K.—a very different picture. The energy output/input ratio and energy use per unit of crop land is even more unbalanced with values of 0.87 (U.S.) vs 0.34 (U.K.) and 12.9 $GJ \cdot ha^{-1}$ (U.S.) vs 49.4 $GJ \cdot ha^{-1}$ (U.K.), respectively.

This indicates that United States agriculture is not among the most energy intensive food production systems of the world. Due to greater inputs per hectare in the United Kingdom agriculture, the yields are also greater. This indicates that United States agriculture could obtain greater yields through the use of increased inputs, but at the expense of lower production efficiencies. Presenty, the United States is operating its agriculture on less energy per hectare than many other industrialized food systems. This is possible because we have more arable land per capita and have not been forced to maximum yields. However, as population pressures increase, it can be expected that energy input per hectare will increase even while the cost of energy is rising.

Countries with industrialized food systems that are importing a large portion of their food may be near their maximum yield or well into the area of diminishing returns and could be expected to have only small increases in energy inputs. On the other hand, food production systems that use less energy per hectare than the United States, primarily developing countries, can be expected to show significant increases in energy use. While high energy productivity is desirable, it is usually linked with low agricultural productivity. Conversely, high agricultural productivity is energy-demanding. It may appear that energy inputs resulting from mechanization (tractors and equipment) would not necessarily increase yields; however, they normally do since critical operations can be accomplished in less time thus avoiding adverse weather conditions.

Table 3.8 gives a comparison of energy use in food production for the United States, the United Kingdom and India.

Plantier (1977), in discussing energy use in European agriculture, indicated that total energy needs of agriculture have more than doubled in the last 10 years. This added energy has led to increased output per unit of land but also an increase in energy consumption per unit of output. In the United States, increased consumption of energy has led to far higher yields, but the energy required per unit of output (energy productivity) has changed only slightly.

The distribution of energy use among various energy-requiring inputs to agriculture varies considerably from one country to another. In Sweden (Plantier 1977), fuel and tractors accounted for about 43% and fertilizer for some 37% of energy consumption in agriculture in 1972, figures that are comparable to United States agriculture. In Yugoslavia, transport accounts for about 60% of the energy used in agriculture. The enlargement and merging of farms and the development of cooperatives are helping to reduce this transport and bring about more rational use of equipment. In the Netherlands, heating of greenhouses accounts for about 79% of all energy directly consumed in the agricultural sector (Dekkers *et al.* 1974).

The distribution of energy consumption among sectors within the food chain also varies from country to country. For example, energy consumption in production agriculture is approximately the same percentage (3 to 4%) of the total energy consumption in Europe, as it is in the United States. However, in the remainder of the food chain, the percentage for the United States is about double that for the European countries.

Another factor that varies widely from country to country is the annual energy consumption per farm worker. These values were 25.2 GJ for the United States, 14.4 in Germany, 7.6 for the U.S.S.R. and 1.1 in Italy. Of course these figures are the direct result of two factors which generally

TABLE 3.8. ENERGY AND FOOD PRODUCTION, 1970

	United States	United Kingdom	India[1]
Total Fossil Fuel Energy ($mg \cdot 10^{12} \cdot PJ \cdot yr^{-1}$) Put Into:			
Primary crop production	1543	227	100
Primary agricultural production	2204	396	100
Total food system	9842	1965	?
Total nation for all purposes	71320	8618	?
Fossil Fuel Energy ($GJ \cdot yr^{-1}$) Put Into:			
Primary crop production	7.6	4.1	0.2
Primary agricultural production	10.8	7.1	0.2
Total food system	48.2	35.2	?
Total nation for all purposes	349	154	?
Fossil Fuel Energy Per Hectare ($GJ \cdot yr^{-1}$) Put Into:			
Primary crop production (cropland only)	12.9	47.2	1.0
Primary agricultural production (total agricultural land)	5.0	20.0	0.6
Total Food Energy ($mg \cdot 10^{12} \cdot PJ \cdot yr^{-1}$)			
Produced at farm gate	1911	135	?
Wasted in processing and distribution	191	13	?
Supplied to consumers	1720[2]	122	?
Eaten or wasted by total population	973[2]	276	1700
Eaten or wasted per person	13.07	13.27	8.5

Source: Green (1978).
[1]The figures for India are very approximate because no reliable sources of information have been located.
[2]From these figures it appears that the United States has $747\ MJ \cdot 10^{12} \cdot yr^{-1}$ of food energy available for export or enough to give 170 million people $12\ MJ \cdot day^{-1}$ if none were lost in processing, transport or distribution.

vary in opposite directions: energy consumption and amount of labor on the farm. Table 3.9 shows the percentage of the laboring population in agriculture and the average percentage of income spent on food.

As population pressures increase, it will become more important for many countries to become self-supporting in food production because food for import may not be available. The Netherlands appears to have accomplished this goal even though it is one of the most densely populated areas of the world. It is estimated that the potential primary production is still about twice the present level. It appears that much progress could be made in making more countries food self-supporting if a worldwide emphasis was placed on improved food production systems.

TABLE 3.9. AGRICULTURAL MANPOWER AND PORTION OF INCOME SPENT ON FOOD

Country	Percentage of laboring population working in agriculture	Average percentage on income spent on food
United States	7	18
Canada	11	21
Australia	11	22
Denmark	17	22
South Africa	30	28
Ireland	35	33
U.S.S.R.	39	56
India	73	66
Nigeria	97	70

Source: FAO statistics *in* Green (1978).

BIBLIOGRAPHY

BLAXTER, K.L. 1975. The energetics of British agriculture. J. Sci. Food Agric. *26* (8) 1055–1064.

BROWN, S.J. and BATTY, J.C. 1976. Energy allocation in the food system: A microsale view. Trans. ASAE. *199* (4) 758–761.

COUNC. FOR AGRIC. SCI. AND TECHNOL. (CAST). 1977. Energy use in agriculture. Report No. *68.* Counc. for Agric. Sci. and Technol., Agrcn. Dep., Iowa State Univ., Ames.

DEKKERS, W.A., LANGE, J.M. and DEWIT, C.T. 1974. Energy production and use in Dutch agriculture. Neth. J. Agric. Sci. *22* (2) 107–118.

DORNOM, H. and TRIBE, D.E. 1976. Energetics of dairying in Gippsland. Search *7* (10) 431–433.

ECON. RES. SERV. (ERS). 1974. The U.S. food and fiber sector: Energy use and outlook. U.S. Dep. Agric., Washington, D.C.

FED. ENERGY ADMIN. (FEA). 1976. Energy use in the food system. Fed. Energy Admin., Washington, D.C.

GIFFORD, R.M. and MILLINGTON, R.J. 1975. Energetics of agriculture and food production. Bull. No. *288.* Commonwealth Sci. and Ind. Res. Organ. (CSIRO), Australia.

GREEN, M.G. 1978. Eating Oil—Energy Use in Food Production. Westview Press, Boulder, Colorado.

HANDRECK, K.A. and MARTIN, A.E. 1976. Energetics of the wheat/sheep farming system in two areas of South Australia. Search *7* (10) 436–443.

HAWTHORN, J. 1975. Energy usage in food processing and distribution. Span *18* (1) 15–16.

HEADY, E.O. 1976. The Agriculture of the United States. Sci. Amer. *235* (3) 106–127.

HEICHEL, G.H. 1976. Agricultural production and energy resources. Amer. Sci. *64* (1) 64–72.

HIRST, E. 1973. Energy use for food in the U.S., *ORNL-NSF-EP-57.* Oak Ridge Natl. Lab., Oak Ridge, Tenn.

HIRST, E. 1974. Energy for food: From farm to home. Trans. ASAE *17* (2) 323–326.

INST. FOOD TECHNOL. 1977. Overview-energy analysis. Food Technol. *21* (3) 51–87.

KLEIS, R.W., *et al.* 1976. Energy related impacts on great plains agricultural productivity in the next quarter century, 1976–2000. Pub. No. *82,* Univ. of Nebraska, Lincoln.

LEACH, G. 1976. Energy and Food Production. IPC Science and Technology Press, Guildford, Surrey, U.K.

NEWCOMBE, K. 1976A. Energy use in the Hong Kong food system. Agro-Ecosystems *2* (4) 253–276.

NEWCOMBE, K. 1976B. The energetics of vegetable production in Asia, old and new. Search *7* (10) 423–430.

PEARSON, R.G. and CORBET, P.S. 1976. Energy in New Zealand agriculture. Search *7* (10) 418–423.

PIEROTTI, A., KEELER, A.G. and FRITSCH, A.J. 1977. Energy and food. CSPI Energy Series *X.* Cntr. for Sci. in the Public Interest, Washington, D.C.

PIMENTEL, D. *et al.* 1973. Food production and the energy crisis. Science *182* 443–448.

PLANTIER, R. 1977. The use of energy in European agriculture. Monthly Bull. of Agric. Econ. and Stat. *26* (6) 1–9.

RICE, R.A. 1972. System energy and future transportation. Technol. Rev. *74,* 31–37.

SINGH, R.P. 1978. Energy accounting in food processing operations. Food Technol. *32* (4) 40–44, 46.

SINGH, G. and CHANCELLOR, W. 1975. Energy inputs and agricultural production under various regimes of mechanization in northern India. Trans. ASAE *18* (2) 252–259.

SMERDON, E.T. *et al.* 1975. Fuel use estimates for Florida agricultural production. Inst. Food and Agric. Sci., Univ. of Florida, Gainesville.

STANHILL, G. 1974. Energy and agriculture—A national case study. Agro-Ecosystems *1* (3) 205–217.

STANSFIELD, J.R. 1975. Fuel and power in agriculture. Span *18* (1) 23–24.

STEINHART, J.S. and STEINHART, C.E. 1974. Energy use in the U.S. food system. Science *184,* 307–316.

STICKLER, F.C., BURROWS, W.C. and NELSON, L.F. 1975. Energy, from sun, to plant, to man. Deere & Company, Moline, Ill.

TUCKER, V.A. 1975. The energetic cost of moving about. Amer. Sci. *63* (4) 413–420.

UNGER, S.G. 1975. Energy utilization in the leading energy-consuming food processing industries. Food Technol. *29* (12) 33–45.

U.S. DEP. AGRIC.—FED. ENERGY ADMIN. (USDA—FEA). 1977. A guide to energy savings for the orchard grower. U.S. Dep. Agric., Washington, D.C.

WATT, B.K. and MERRILL, A.L. 1973. Composition of foods: raw, processed, prepared. Agric. Handbk. No. *8,* U.S. Dep. of Agric., Washington, D.C.

4

Agricultural Energy Analysis

ENERGY ANALYSIS

Energy analysis (EA) is a young and fast developing discipline, parallel in some ways to economics, yet with distinct differences. Beardsworth (1975) stated that energy analysis is the objective analysis of the physical quantities of energy involved in a process, system, etc. That is, the identification and measurement of energy flows constitute energy analysis.

Energy is a pervasive quantity, at least as pervasive as any other input for producing a good or service. Energy is essential to all activity. No action will occur within a system without an enabling energy flow and a system-wide entropy increase. Energy prices are increasing and resources of nonrenewable energy are being depleted. All of these factors point to the importance of energy analysis as an emerging discipline.

The discipline of energy analysis is characterized by many different ideas, procedures, metholodogies and approaches. For example, Slesser (1977) characterized two distinct "schools" of energy analysis thought: the Odum school and the IFIAS (International Federation of Institutes for Advanced Study) school. Hoffman (1975) characterized three approaches to energy analysis: those of economics, engineering or process analysis, and net energy. Hill and Walford (1975) listed as many as six forms of energy analysis.

Table 4.1 shows the characteristics of the two schools identified by Slesser (1977). As can be seen in the table, the stated purposes of these two schools vary. The IFIAS school (IFIAS 1974,1975) lists as its purposes the facilitation of conservation, forecasting demand and affecting demand, predicting the effects of shortages, and understanding

TABLE 4.1. TWO SCHOOLS OF ENERGY ANALYSIS

Characteristic	Odum School (Eco-energetic)	IFIAS School (Sequestered)
Definition of energy analysis	Modeling of systems accompanied by an evaluation of energy flows.	Determination of how much energy is sequestered in a good or service.
Scope	Interdependency and linkages between man's fuel powered systems and the natural ecosystems.	Quantities of energy resources consumed and, therefore, resource depletion.
Energy sources	Sun and labor as well as nonrenewable energy resources.	Nonrenewable energy resources.
Energy quality	Provision made for different energy qualities via use of fossil fuel equivalents.	Provision via summation of useful and wasted energy in producing all inputs required for a good or service.
Energy theory of value	Accepted by some analysts.	Rejected. A multiple factor theory of value, including such as labor, land and capital, can better treat processes which involve more than a single factor.
Energy content of labor	Human service termed a high quality energy flow.	Minimized importance of labor energy in industrialized economies.

system thermodynamics (Chapman 1974A). Bayley *et al.* (1976) states for the Odum school that energy analysis has "evolved into an energy evaluation procedure which can be used to assess resource management alternatives."

It is likely that as the discipline of energy analysis grows and matures, differences and distinctions between the several groups and schools of energy analysis will disappear. Conventions will be agreed upon and the groups will unite. Evidence of development and evolution of thought is readily seen by comparing Odum (1971) with Odum and Odum (1976). Beardsworth (1975) suggested that thermodynamics might act as a unifier or integrator of "economics, physical technical analyses and perhaps matters involving the biosphere." Georgescu-Roegen (1971,1975) has espoused this position for several years.

Several methods are used for determining the energy sequestered in goods and services or for performing an energy analysis on a good or service. Chapman (1974A) and Pimentel *et al.* (1974) reviewed the methods:

(1) *Statistical analysis:* Determining the energy sequestered per unit of output from statistical data. For instance, the U.S. Bureau of Census (1976) reported a 1975 U.S. energy consumption of $71{,}078 \times 10^{12}$ Btu (fossil fuels, hydro and nuclear) and a GNP (gross national product) of $\$1516 \times 10^{9}$. Therefore, the average dollar's worth of goods and services produced consumed 47,291 Btu or 49.89 MJ or 11,917 kcal.
(2) *Input-output analysis:* A square matrix of (typically) a national economy presents the quantities of each commodity (in quantities, energy units or monetary units) required to produce every commodity. Material flows can be traced backwards to eventually the primary energies upon which they depend.
(3) *Process analysis:* The networks or processes required to make a final product are identified. Each is analyzed to determine its inputs. Each input is assigned an energy requirement so that the total energy requirement can be summed.

Both the latter two methods of analysis are widely used when better estimates are desired than can be obtained through statistical analysis. Each of these methods will be described later in greater depth. Bullard *et al.* (1976) showed how to combine process analysis and input-output analysis to obtain the best estimate of energy intensity or embodied energy with least effort. They indicated that input-output analysis is better for aggregated, nationwide problems and that process analysis is more suited to specific processes, products or manufacturing chains for which physical flows of goods and services are easy to trace.

One particular problem in energy analyses is that of nonhomogeneity of different energy forms or sources (Webb and Pearce 1975; Leach 1975; Hill and Walford 1975; Huettner 1976). This prevents, or makes subject to error, simple additions of energy flows from different sources. Two fuels might have the same heat or energy content, say a MJ of coal and a MJ of electric power, but have other significantly different attributes. Their economic values, entropy, uses, quality, ability to do work, concentration, cleanliness, etc., may differ. Odum and Odum (1976) provided one conceptual mechanism to accomodate these differences and achieve comparability with their suggested fossil fuel equivalents (see list of definitions in Appendix). These values are based on the quantities of one energy required to derive another in typical real-world systems. An unanswered question is what value of FFE (fossil fuel equivalents) should be used for an energy form derivable from another by more than one process having different efficiencies.

Another problem in energy analyses is in determining where to draw boundaries around the system or process being studied. Beardsworth

TABLE 4.2 ENERGY EQUIVALENTS

Type of Energy	MJ (MJ of heat to make one FFE MJ)	Fossil Fuel Equivalents (FFE MJ · heat MJ^{-1})
Heat from sun's rays, uncollected	10,000	0.0001
Sunlight	2,000	0.0005
Gross plant production	20	0.05
Wood, collected	2	0.5
Coal and oil delivered for use	1	1.0
Energy in elevated water	0.33	3
Electricity	0.25	4

Source: From *Energy Basis for Man and Nature*, Odum and Odum (1976).

(1975) called this the problem of "attributability"; that is, which energy inputs should be included or attributed as energy "costs" to a particular output and which should be excluded. Leach (1975) described another aspect of the boundary problem: The determination of which *outputs* should be included. For instance, for an energy production sector, the boundaries can be drawn sufficiently large so that some or many of the inputs to that sector can be within the boundary and therefore not included as outputs. In other words, using the example of owning and operating an automobile, does one include as useful output the mileage the automobile is driven to have it fueled, serviced, repaired and inspected?

A third problem in energy analyses is that of multiple or joint outputs. When there is more than a single output from a process or system, how is the assignment of energy inputs divided among the outputs? Chapman (1974B) examined allocation by weights, prices and enthalpies of output products. He favored the latter. Leach (1975) said none of these avoided logical absurdities when the outputs include a fuel and a nonfuel which have zero substitutability for each other.

Finally, it should be emphasized that energy is, despite its pervasiveness, not the only input for the production of goods and services. A theory of value based upon energy is incomplete. Other inputs which may not be properly and adequately described in energy terms are required, such as land, labor and capital. These various inputs are usually partially substitutable one for another but usually not completely so. Hoffman (1975) stated that "some single 'numeraire' is needed to sum up the inputs to determine the 'value' of a product," then pointed out that monetary units have been used, but that these and any other single numeraire including an energy equivalent causes information to be lost in the transformation and implies a substitution possibility that may not exist. An energy theory of value, despite its perhaps initial attractiveness, seems to be useless.

Energy analyses have led to a quantification of the energy sequestered in many items. Leach and Slesser (1973) provided a process energy analysis of the energy sequestered in various inputs to agricultural production, including fuels, fertilizers and transportation. Their values have been widely used in subsequent literature.

Makhijani and Lichtenberg (1972) provided energy requirements for producing and processing many raw materials represented in Table 4.3. A more recent source of such information is represented in Table 4.4.

CONCEPTS AND TOOLS FOR AGRICULTURAL ENERGY ANALYSIS

Agricultural energetics includes the identification, measurement and analysis of energy flows in agricultural systems. We have seen examples of the energy flows in a number of different types of agricultural systems from primitive to industrialized and also the energy flows associated with food in natural ecosystems and in hunting-gathering societies. Now we will attempt to further develop the concepts and tools necessary for agricultural energy analysis.

In order to evaluate the energetics of agricultural and food systems, we need measures of how well such systems use energy. Energy ratios have been used extensively and will be discussed first. A new measure, energy productivity, will also be explained. Each has an application but generally for different systems.

TABLE 4.3. ENERGY REQUIREMENTS FOR RAW MATERIALS

Basic Material	Energy Intensity ($MJ \cdot kg^{-1}$)
Aluminum	262
Steel	46.4
Paper	23.4
Plastics	9.52
Lumber	2.5
Sand and gravel	0.071

Source: Makhijani and Lichtenberg (1972).

TABLE 4.4. PRIMARY ENERGY CONSUMPTION FOR UNITED STATES PRODUCED PRODUCTS IN 1970

Product	Primary Energy (Energy Intensity) ($MJ \cdot kg^{-1}$)	Primary Energy Cost (% of selling price)
Low density polyethylene resin	108.7	15.4
High density polyethylene resin	103.1	14.5
Polystyrene resin	136.5	30.3
Polyvinyl chloride resin	96.4	12.0
Portland cement, wet process	9.3	19.3
Portland cement, dry process	8.4	16.6
Primary copper	130.1	3.8
Primary aluminum	201.5	9.8
Raw steel	22.4	10.8
Glass containers	21.1	4.6
Newsprint	25.5	5.3
Writing paper	28.5	4.5
Corrugated containers	25.3	4.9
Folding boxboard	25.5	5.0
Virgin styrene butadiene rubber	155.4	15.0

Source: FEA (1975).

Energy Ratio

Scientists have long used energy ratios to evaluate the energetics of primitive and subsistence societies. Rappaport (1976) described the ratio of the energy contained in the food grown to the human energy expenditure of the Tsembaga's, a population of bush-fallowing horticulturists of New Guinea. In such analyses where there are few if any flows of primary energy from outside the society, it is useful and logical to examine agriculture's energy production.

In both hunting-gathering societies and subsistence primitive agricultural societies, the energy contained in the output food products must be sufficient to supply people with energy to continue functioning as hunters, gatherers or farmers. If they do not produce sufficient food within their social groups (families, tribes, villages), they go hungry. Therefore, the energy ratio must be sufficiently large. However, a 1:1 energy ratio is not necessarily adequate. It instead appears that a ratio in the order of 15:1 to 20:1 must be achieved to insure adequate food. Norman (1978) has followed Black's (1971) convention in using total crop energy output relative to total net energy expended in production tasks. This net energy expended was taken by Black to be 0.63 $MJ \cdot h^{-1}$ and by Norman as 0.75 $MJ \cdot h^{-1}$. This is the additional energy expended above what would otherwise have been expended had the subject been sedentary, making the total gross energy expenditure about 1.00 $MJ \cdot h^{-1}$ when working. If it is assumed that the laborer works 4–5 $h \cdot d^{-1}$, then the total energy per worker is about 10–12 $MJ \cdot d^{-1}$ (1.00 $MJ \cdot h^{-1} \times 5$ h +0.25 $MJ \cdot h^{-1} \times 19$ h = 9.75 $MJ \cdot d^{-1}$), and the worker expends about 30–50% of his/her total gross energy expenditure in field work. Norman further suggests that most family members, unless very young, aged or sick, work in the field; but because there are those young, aged, or sick who do not, the ratio of gross energy expended in crop production to total gross energy expended is in the order of 0.2 to 0.4. Also, crop yields are sometimes less than average and there are also storage, processing and household waste losses. His conclusion is that energy ratios must be in the order of 15:1 or above to allow long-term survival of the society. Figures 4.1 and 4.2 from Leach (1976) show energy ratios for both subsistence primitive agricultural and industrialized agricultural societies.

Black (1971) was one of the first to apply energy ratio, which he termed as efficiency ratio (E), to industrialized agricultural production systems. Black, although noting certain indirect energy inputs, limited energy inputs to the fuel consumed by tractors. His conclusion, based upon his thus simplified analysis, was that industrialized agriculture was no more efficient in energy use than hand or draft animal cultivation.

Our complex society with its interdependent networks of individuals, skills, industries, resource suppliers and political boundaries has not developed into a society of subsistence and independence, but rather as one of dependency as each tends to do the thing or produce the goods and services for which he/she is best suited. In such a setting, if there are sufficient external sources of energy, there is no necessity that a farmer be energy independent. Therefore, in such a setting, energy ratio has minimal meaning and use.

Nevertheless, further applications were quickly made of energy ratio in industrialized agriculture. Perelman (1972) compared the United States food production to Chinese wet rice agriculture, emphasizing the much greater efficiency of the latter according to the compared energy ratios. Pimentel *et al.* (1973) expressed the results of an analysis of the United States corn production as kilocalories of corn per kilocalorie of energy inputs. In Hirst's (1973) analysis of the United States food system, the reciprocal, energy use per unit of food energy, was termed energy ratio. Heichel (1973) defined the ratio of food energy to input energy as "caloric gain"; he specifically excluded sunlight as a component of total input

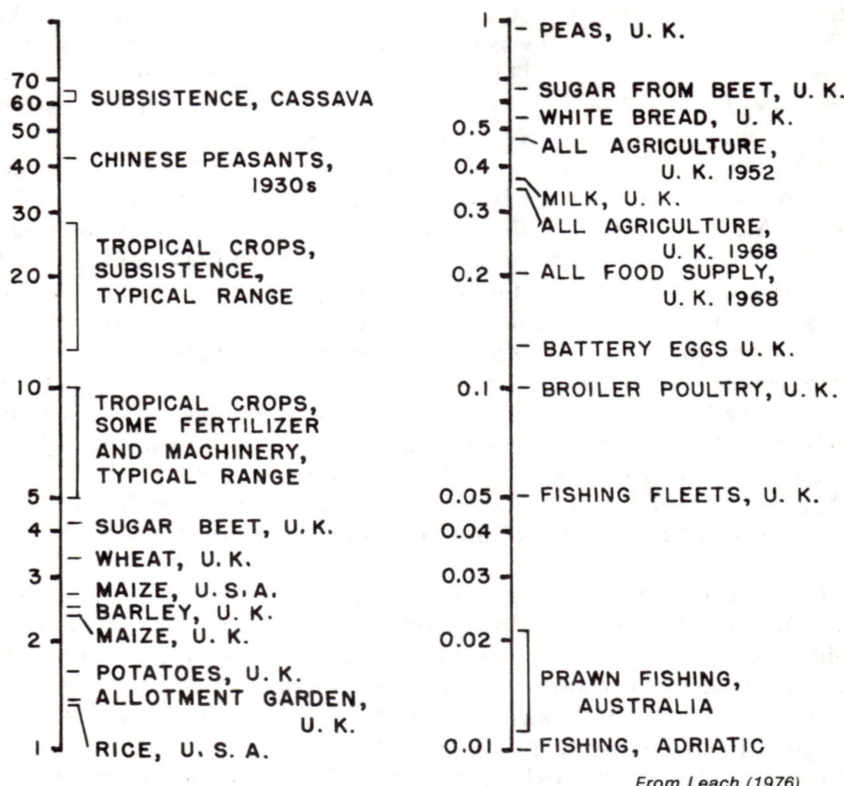

FIG. 4.1. ENERGY RATIOS FOR FOOD PRODUCTION

energy and termed the total input energy "cultural energy." Steinhart and Steinhart (1974) termed the input "energy subsidy" and also used the reciprocal of energy ratio to show that for the United States the input energy to the food system has increased sharply relative to the food energy output in this century. Leach (1976) maintained the infant tradition of using energy ratio to evaluate the energetics of all types of agricultural systems in his research.

It has further been recognized that foods are consumed for reasons other than their energy content. Protein has also been identified as the output in an energy context. Hirst (1973) and Hannon *et al.* (1976) used the ratio of input energy to protein to compare several categories of food products. Heichel (1976) used protein output as the numerator with energy input as the denominator. Rawitscher and Mayer (1977) evaluated the energy requirements of fishing for several seafoods on the basis of energy per unit of protein.

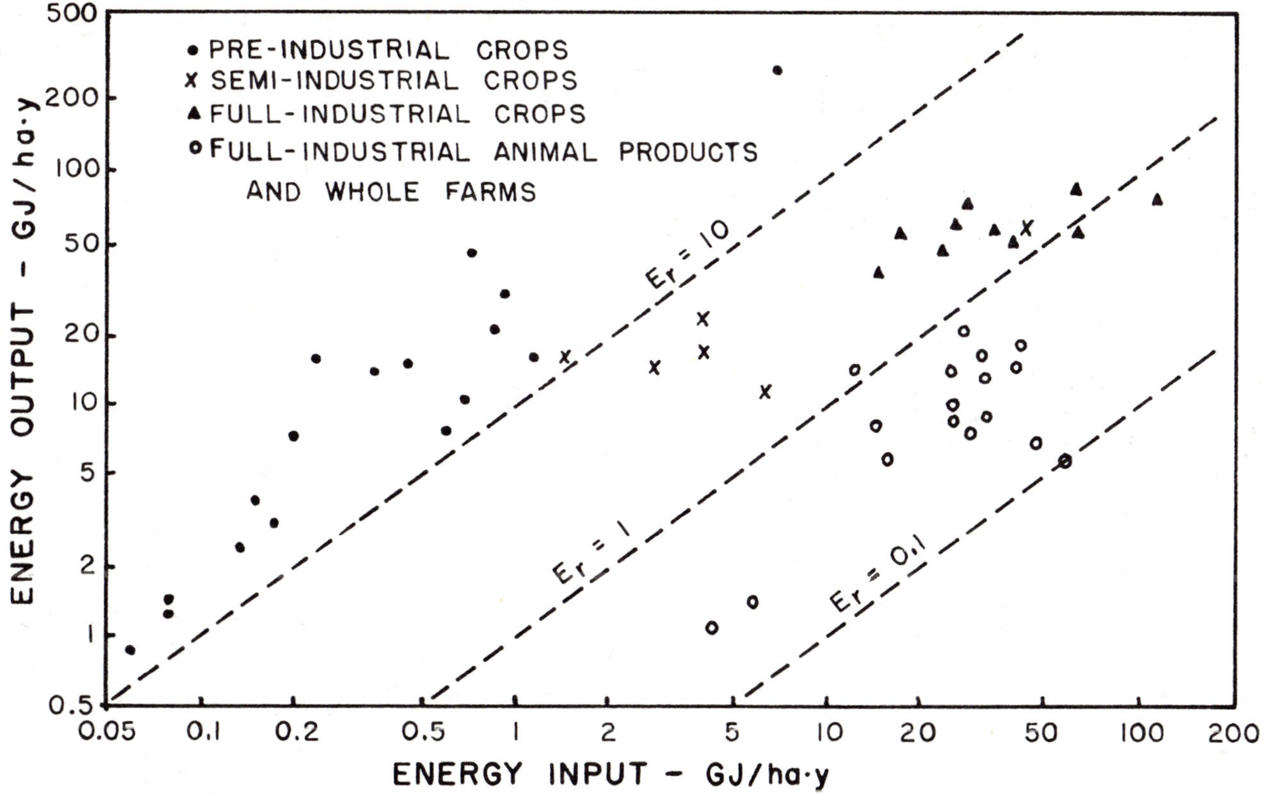

From Leach (1976)

FIG. 4.2 ENERGY INPUTS AND OUTPUTS PER UNIT LAND AREA IN WORLDWIDE EXAMPLES OF FOOD PRODUCTION

The fact shown is that food has more than one attribute which leads to the first limitation of energy ratios and similar evaluations. Comparison of the input energy to one or more of the qualities of the food produced must always be incomplete. Hill and Erickson (1975) persuasively argued: "The fallacy of comparing countries, diets, or food items on the basis of calories illustrates a fact that everyone recognizes with a moment's reflection. In developed countries, food is a form of entertainment, business function, a social amenity, a psychological escape, or a form of peer acceptance. Only rarely is it used *primarily* as a source of nourishment to meet biological requirements." Gifford (1976) and Doering and Peart (1977) also recognized this limitation of energy ratio.

A second limitation of energy ratios, identified by Gifford (1976), is the incorrect conclusion or at least inference that in industrialized agriculture, a ratio of less than unity is unacceptable. DeWit (1975B) stated that ". . . farming for energy must be energetically sound, but there is no reason why farming for other products should cost less energy than is produced." Spedding and Walsingham (1975) agreed, stating that energy ratios ". . . are only strictly relevant to agricultural systems intended to produce energy."

Odum and Odum (1976) explained the theory of net energy; they showed that high quality energy flows (fossil fuels) may interact with low quality energy flows (sunlight) to maximize the total flow of energy. The low quality energy flows do not have to yield net energy in isolation. It is necessary instead that the total energy flow be increased by the addition of low quality energy flow to the high quality energy flow. Therefore, for industrialized agriculture, it is not necessary that there be net energy or that the energy ratio be greater than unity. The single important exception are systems designed to produce energy, for instance, "energy plantations," rather than food or another agricultural product.

A third limitation of energy ratios lies in the manner in which they have been used. Comparisons have been made of the energy ratios of different agricultural products and of different cultures or agricultural systems. Such comparisons are meaningless and misleading. The food-fossil energy conversion ratio varies according to product and cultural practices (Breimyer 1975). Further, although an energy ratio could be calculated for agriculturally produced nonfoods such as fibers and ornamentals, it would have little rational basis (DeWit 1975B).

The need for an improved measure of the use of energy in agriculture has been recognized by the critics of energy ratio. Doering (1974) stated ". . . the exclusive use of caloric input-output ratios is both insufficient and misleading. A much broader range of criteria needs to be established before a narrow accounting system is foisted upon us." Doering and Peart (1977) pointed out the need to know energy consumption of an agricultural system only in comparison to the productivity of that system. DeWit (1975A) said that government policy should be directed towards a decrease of energy use per unit of product.

Several reports have related the quantity of production to the energy inputs of agricultural production systems. Slesser (1973) compared the productivity of a large number of agricultural systems and products in kilocalories per kilogram. Lockeretz (1975) compared several beef production systems on the basis of the millions of Btu required per hundred pounds of separable lean beef. Cervinka *et al.* (1975) quantified the gallons of fuel required to produce a tonne of several different vegetables in California. Makhijani (1975) compared rice, wheat and maize production of six subsistence agricultural systems in millions of Btu of energy inputs per ton product. Roller *et al.* (1975) synthesized several beef production systems and compared them on the basis of megacalories per kilogram of live weight. Avlani and Chancellor (1977) evaluated the energy requirements of California wheat production in thousands of Btu per ton. Spedding and Walsingham (1975) suggested that for some products, wool for example, the energy cost per unit of wool produced would be relevant. Pierotti *et al.* (1977) determined the energy requirements for producing a number of crop and livestock products in kilowatt hours per weight or volume unit of production. These examples relating production to energy inputs indicate a need for such a generalized measure. This need will be addressed in the next section.

Energy Productivity

Agricultural production systems utilize numerous inputs in providing output products for human consumption. Among the ones most widely recognized are land, labor, knowledge, capital, natural resources, water, sunlight, minerals and energy. The costs of these inputs vary. The successful manager combines varying quantities of available inputs to produce products with low total cost.

Productivity is the measure of how much product is obtained per unit of input. Land productivity is a universally important measure, so much so that "productivity" is often used when "land productivity" is intended. A second productivity measure of high importance in many industrialized countries is labor productivity, the quantity of output per unit of labor input. The importance of labor productivity directly results from our high standard of living and the importance we place upon the individual. Other measures of productivity have been used (Comm. Agric. Prod. Efficiency 1975) depending upon their cost and importance in affecting yields.

Energy is costly and likely to become more so. Fossil fuel energy is yet relatively abundant but will become scarce. It is of utmost importance that our agricultural production systems utilize energy as efficiently as possible, consistent with other goals and constraints.

It has therefore been proposed by Fluck (1979) that a new measure of productivity, the quantity of product per unit of input energy, be designated and that it be termed energy productivity. In the SI system of units, a convenient measure of energy productivity is kilograms per megajoule ($kg \cdot MJ^{-1}$). Breimyer (1977) also proposed a similar measurement.

Energy productivity is specific for each agricultural product, location and time. That is, energy productivity can be used only to compare alternate production systems and energy conservation practices which result in the same product, at the same place, at the same time. End points for comparable systems must be identical. Energy productivity is not intended to evaluate different products. Consumers seldom, if ever, select products to purchase based upon the energy sequestered in the products. Energy productivity *is* intended to and can serve as an evaluator of how efficiently energy is utilized in production systems yielding a particular product.

Several characteristics of energy productivity should be noted:

(1) Energy productivity is dependent upon where in the total system the end point is set. As more of the total system is included, energy productivity is lowered. For example, more energy is required to deliver a given quantity of a product to a supermarket than is required to deliver it to the farm gate, and hence, the energy productivity is less at the supermarket. This is true even though the product is essentially unchanged.

(2) Energy productivity is less as losses occur, resulting in less product for a given quantity of sequestered energy.

(3) If omissions occur in tabulating the energy inputs in a production system, the calculated energy productivity will be incorrectly greater than it should be.

(4) Energy productivity of livestock and livestock products is lower than that of feed inputs to their production due to the conversion inefficiencies and additional livestock system energy inputs.

This proposal for a new measure of the use of energy in agricultural systems does not necessarily alleviate two additional objections to energy ratio raised by Gifford (1976). One objection was to the simple addition of different forms of energy in determining total inputs such as the heat contents of fossil fuels, hydroelectric power, nuclear power and muscle power. This objection potentially has several components. First, the energy content of electric power from nonfossil fuel sources has historically been determined by the United States Bureau of Mines as though the electric power were generated from fossil fuels. This practice measures the depletion of fossil fuels and their potential depletion if these nonfossil fuel alternatives were not in use. Second, the energy of muscle power should not be measured as such but also in terms of the fossil fuels depleted (Fluck 1979). Third, heat contents from various fossil fuels are not treated differently despite the fact that, for many uses, some fossil fuels are preferable to others. Therefore, two of the methodological objections are easily alleviated, but the third remains.

Gifford's other objection was that to study energy inputs out of context from other inputs substitutable for energy has little meaning. To this we agree and in no way suggest otherwise.

Energy productivity values calculated from available data for industrialized agricultural systems are given in Tables 4.5 and 4.6. Values of energy productivity range from only slightly greater than 0.01 kg · MJ^{-1} for certain agricultural products to less than 2 kg · MJ^{-1} for others. These examples are given only to illustrate typical values. They are not intended to be used to compare production systems, as the products are different.

Fluck (1975) estimated the energy required for 20 commercial vegetables grown in Florida. Totals were determined for production through retailing. Energy productivities (kg · MJ^{-1}) ranged from 0.0102 for strawberries to 0.0473 for watermelons with sweet corn at 0.0274, cabbage at 0.0457, potatoes at 0.0425, and tomatoes at 0.0170.

Readers are cautioned about using Federal Energy Administration (FEA 1976) energy data to directly calculate energy productivities. Since the only indirect energy inputs included are fertilizers and pesticides, calculated energy productivity values for crops would be biased somewhat upward. For livestock, since the energy sequestered in feeds is not included, calculated energy productivity values would be biased upward considerably.

Energy ratios and similar measures dependent upon the energy content of the product have been shown to be inappropriate as measures of the efficiency of energy utilization in industrialized agricultural systems. Energy productivity, the quantity of product per unit of energy required for its production, is proposed as an improved substitute. The characteristics and limitations of this new measure of the efficiency of energy utilization in agricultural systems have been identified. Literature reflects dissatisfaction with energy ratio and a focus on energy productivity. The time seems right for its adoption. It is so recommended.

It is cautioned that energy productivity should not be taken as the sole criterion of efficiency in any general sense. Other inputs are also scarce

TABLE 4.5. SELECTED ENERGY PRODUCTIVITIES OF UNITED STATES GROWN CROPS AND LIVESTOCK

Product	Energy Productivity ($kg \cdot MJ^{-1}$)
Crops:	
Corn	0.1846
Sorghum	0.1155
Rice	0.0819
Oats	0.2822
Barley	0.1819
Rye	0.1922
Wheat	0.1965
Other hay	0.4251
Alfalfa	0.3332
Corn (silage)	1.0053
Cotton	0.0137
Tobacco	0.0150
Soybeans	0.1893
Potatoes	0.3762
Vegetables	0.346
Edible beans, dried	0.0675
Sweet Potatoes	0.365
Edible peas, dried	0.307
Grapes	0.2299
Apples	0.1609
Peaches	0.1531
Peanuts	0.1014
Flaxseed	0.1233
Livestock Products:	
Cattle on feed	*0.0110
Other beef cattle	*0.0216
Hogs	*0.0216
Sheep and lambs	*0.0212
Chickens	*0.0341
Turkeys	*0.0201
Eggs	0.0301
Milk	0.1364
Lard	0.0653
Edible tallow	0.1054

Source: Pierotti *et al.* (1977).
*Kilograms of meat produced per megajoule of energy

(land, labor, water, etc.) and the efficient use of one may conflict with another, i.e., tradeoffs are involved and most viable agricultural production systems tend to optimize the use of the various production inputs available.

Input-Output (I/O) Analysis

Input-output analysis relates an industry's inputs to its outputs and to the inputs and outputs of other industries with which it interacts. The entire economy of a country, region, etc., is divided into a finite number of industries or sectors, and all transactions of goods and services

TABLE 4.6. SELECTED ENERGY PRODUCTIVITES OF UNITED KINGDOM GROWN CROPS AND LIVESTOCK

Product	Energy Productivity ($kg \cdot MJ^{-1}$)
Crops:	
Wheat flour	0.0686
Barley	0.0767
Oatmeal	0.0594
Potatoes	0.315
Beet root	0.533
Carrots	1.052
Parsnips	0.514
Turnips and swedes	1.414
Onions	1.01
Brussel sprouts	0.737
Cabbage	0.854
Cauliflower	0.967
Peas (shelled)	0.378
Beans, broad	0.3
Beans, runner and French	1.52
Other vegetables	1.71
Fruit	0.25
Livestock Products:	
Cattle, meat	0.0763
Sheep and lamb, meat	0.0721
Pigs, meat	0.0585
Eggs	0.151
Milk	0.368
Butter	0.0323
Cheese	0.0580

Source: Leach (1976).

between sectors are quantified so that the inputs to each sector are also outputs from the other sectors which support it with raw materials, services, etc. Transactions between sectors are often measured in monetary units, although physical units are also used. A table, or matrix, showing each sector as both a row and column heading, indicates the quantity of product which is the output from one sector (to the left) serving as input to another sector (above). Flows generally occur from raw material industries toward final consumption.

Carter (1974) said: "Input-output coefficients are obtained by dividing the entries in a column, which are an industry's inputs, by that industry's output. In other words, coefficients show the amounts that an industry purchased from all other industries . . . per unit of its own output." For instance, a dollar's worth of tractor (output) might require \$0.34 worth of steel, \$0.03 worth of tires, \$0.01 worth of electrical components, etc.

Carter also described the resulting equations: "Input-output coefficients are used to form a system of linear equations connecting the outputs of all industries. Each equation gives the output, x_i, of a given sector i as the sum of the sector's sales to all other sectors and to final demand, y_i,

$$x_i - \sum_j a_{ij}x_j = y_i$$

where a_{ij} is the input coefficient that tells requirements of the product of sector i per unit of output of sector j. The system has as many equations as there are industries."

Perhaps input-output analysis' greatest advantage is in its ability to determine the effects of some change on all other sectors of an economy, however far removed from the source of the change the other sectors may be. It has been used, for instance, to determine the effects on employment, energy consumption and output in various sectors due to consumers' choosing between mechanical or outdoor clothes-drying or between paper and cloth towels (Herendeen and Sebald 1974).

Several sources (Bullard 1975; Bullard *et al.* 1976; Carter 1974; Chapman 1974A; Herendeen 1973) describe the fundamentals of input-output analysis and applications to energy analysis. For energy analysis, the input-output coefficients indicate the amount of energy a given sector of the economy requires from another sector embodied or sequestered in the product or service transferred between the two sectors. This may be given as the total energy flow per unit time, the energy intensity of the product or service, such as $MJ \cdot \$^{-1}$, or the percentage of the total energy sequestered in the output and supplied by a particular input. The energy flows originate with several primary energy sectors, such as coal mining and crude petroleum and gas, and spread outward to be embodied or sequestered in the output products and services of every sector.

The publications of the Energy Research Group, Center for Advanced Computation, University of Illinois, are sources of basic information on energy analysis utilizing input-output analysis, as well as numerous applications. Herendeen (1973) and Bullard and Herendeen (1975) extensively described their utilization of the U.S. Department of Commerce's 1963 Input-output Table of the United States economy, transformation of it to energy terms and several applications. They determined energy intensity coefficients, total primary energy (the sum of coal, crude oil and gas, and the fossil fuel equivalent of hydro and nuclear) required per dollar of final total demand for a product or service for each of 357 sectors. Tables 4.7 and 4.8 list some of these from an updated source (Bullard *et al.* 1976). These coefficients can be used to estimate the total primary energy sequestered in a product or service based upon its cost.

TABLE 4.7. TOTAL PRIMARY ENERGY REQUIRED PER $1 FINAL OUTPUT FOR AGRICULTURALLY RELATED PRODUCTS

Sector or Industry	Energy Intensity (MJ · 1967^{-1})	(MJ · 1974^{-1})
Agriculture, Forestry and Fisheries:		
Dairy farm products	84.68	41.13
Poultry, eggs	98.71	47.94
Meat animals and misc. livestock products	88.91	43.18
Cotton	128.42	62.37
Food feed grains and grass seeds	87.53	42.51
Tobacco	80.52	39.11
Fruits and tree nuts	55.32	26.87
Vegetables, sugar and misc. crops	54.51	26.47
Oil bearing crops	64.93	31.53
Forest, greenhouse and nursery products	71.22	34.59
Forestry and fishery products	81.89	49.04
Agriculture, forestry and fishery services	45.96	27.52
Manufacturing:		
Meat products	86.49	53.42
Creamery butter	96.82	59.80
Cheese, natural and processed	92.60	57.20
Condensed and evaporated milk	97.31	60.10
Ice cream and frozen desserts	78.13	48.26
Fluid milk	77.06	47.60
Canned and cured sea foods	75.99	46.94
Canned specialties	92.96	57.42
Canned fruits and vegetables	95.17	58.78
Dehydrated food products	83.90	51.82
Pickles, sauces and salad dressings	91.50	56.52
Fresh or frozen packaged fish	77.14	47.65
Frozen fruits and vegetables	93.46	57.73
Flour and cereal preparations	88.26	54.51
Prepared feeds for animals and fowls	104.70	64.67
Rice milling	82.71	51.09
Wet corn milling	160.58	99.19
Bakery products	60.19	37.18
Sugar	149.58	92.39
Confectionery and related products	72.08	44.52
Alcoholic beverages	56.43	34.85
Bottled and canned soft drinks	75.72	46.77
Flavoring extracts and syrups	62.74	38.75
Cottonseed oil mills	149.36	92.25
Soybean oil mills	94.72	58.51
Vegetable oil mills	78.59	48.54
Animal and marine fats and oils	137.73	85.07
Roasted coffee	40.38	24.94
Shortening and cooking oils	111.13	68.64
Manufactured ice	175.08	108.14
Macaroni and spaghetti	74.64	46.10
Food preparations	75.22	46.46

Source: Bullard *et al.* (1976).

TABLE 4.8. TOTAL PRIMARY ENERGY REQUIRED PER $1 FINAL OUTPUT FOR SELECTED INPUTS TO AGRICULTURAL ENTERPRISES

Sector or Industry	Energy Intensity (MJ · 1967$$^{-1}$)	(MJ · 1974$$^{-1}$)
New construction, nonresidential buildings	65.49	33.01
Wooden containers	61.92	35.57
Paperboard containers and boxes	140.75	96.01
Fertilizers	197.34	134.15
Agricultural chemicals	190.33	129.39
Paint and allied products	143.71	91.53
Tires and inner tubes	107.56	77.21
Farm machinery	83.54	59.25
Motor vehicles and parts	81.00	61.31
Motor freight transportation and warehousing	61.35	48.23

Source: Bullard *et al.* (1976).

Table 4.9 shows the distribution of total energy embodied or sequestered in the outputs of two sectors, chemical and fertilizer mining and farm machinery, among the various direct and indirect energy inputs.

Certain limitations of I/O Analysis have been given by Bullard and Herendeen (1975) and Herendeen (1973):

(1) Data accuracy limitations exist due to incomplete coverage of a sector, proprietary information, etc.
(2) Many establishments produce more than a single product or service, but output is attributed to only the predominant good or service.
(3) I/O assumes linearity, i.e., if one dollar's worth of a good requires 30 MJ, then two dollar's worth requires 60 MJ. This is, at best, only an approximation.
(4) Capital goods are not considered as an input but as final demand, which leads to coefficients which are too low.
(5) Coefficients change with time. In fact, the average energy intensity of the GNP has decreased from 92.92 MJ · $$^{-1}$ in 1960 to 49.46 MJ · $$^{-1}$ in 1975.
(6) Demand is measured in producers' and not purchasers' prices.

Process Analysis

Attendants of the first IFIAS energy analysis conference (1974) made several recommendations to standardize the work of energy analysis and in particular to standardize process analysis:

TABLE 4.9. DISTRIBUTION OF SEQUESTERED ENERGY

	Percentage of Total	
	Chemical and fertilizer mining	Farm machinery
Direct Energy:		
Coal mining	0.86	4.65
Refined petroleum	4.36	1.22
Electricity	19.39	3.90
Gas utilities	54.89	5.15
	79.50	14.92
Indirect Energy:		
Iron ore	0.08	
Nonferrous mining	0.18	
Stone and clay mining	0.39	
Maintenance and repair construction	0.24	0.20
Ordnance		0.06
Apparel		0.04
Fabricated textile products		0.01
Wood products		0.18
Wood containers	0.04	0.02
Paper products	0.31	0.08
Paperboard containers	0.02	0.18
Printing, publishing		0.01
Chemical products	3.44	0.14
Paints		0.38
Rubber products	0.39	3.25
Stone and clay products		0.50
Primary iron and steel	2.08	41.77
Primary nonferrous metals	0.12	3.29
Heating and plumbing	0.11	0.36
Screw machine products	0.01	2.73
Fabricated metal products	0.05	1.30
Engines, turbines	0.05	6.10
Construction and mining equipment	0.44	2.08
Materials handling equipment	0.02	0.05
Metal working equipment	0.01	0.59
Special industry machinery		0.15
General industry machinery	0.01	6.02
Machine shop products	0.14	0.66
Service industry machines		0.03
Electrical apparatus	0.05	0.40
Household appliances		0.27
Electric lighting and wiring equipment	0.01	0.01
Misc. electrical equipment		0.75
Motor vehicles and equipment		0.60
Aircraft and parts		0.01
Transport equipment	0.02	0.15
Professional scientific supplies		0.19
Misc. manufacturing	0.01	0.06
Railroad transportation	0.38	0.50
Truck transportation	0.03	0.42
Water transportation	0.01	0.08
Air transportation		0.08
Communications	0.04	0.09
Water and sanitary services	0.03	0.07
Wholesale and retail trade	0.20	1.71

TABLE 4.9. *(Continued)*

	Percentage of Total	
	Chemical and fertilizer mining	Farm machinery
Finance and insurance	0.10	0.23
Real estate	0.32	0.20
Lodging and personal services		0.09
Business services	0.48	1.21
Auto repairing	0.02	0.06
Medical and educational services	0.04	0.04
Federal government enterprises	0.03	0.05
State and local government enterprises		0.01
Business travel	0.31	0.80
Office supplies	0.01	0.09
Imports	10.27	6.04
	20.49	84.89

Source: Bullard *et al.* (1975).

(1) Use "free energy" (see definitions in Appendix A.1) rather than "enthalpy" as the measure of energy available in a fuel.
(2) The amount of an energy source which is sequestered by the process of making a good or service should be the unit of account and it should be termed gross energy requirement (GER) when based on enthalpy and gross free energy requirement (FER) when based on free energy. Some of the GER may be stored in the product or service and made available upon utilization of the product, say, as a waste material.
(3) The GER minus any unconsumed energy recoverable in waste or byproducts should be termed net energy requirement (NER), and the NER should be normally used in assessing the energy requirement of a good or service.
(4) The energy requirement for a process, as only a part of the GER or NER for a good or service, could be determined for comparison purposes, and it should be termed process energy requirement (PER).

Attendants of the conference further recommended a generalized procedure for process energy analysis:

"(1) Decide objective of analysis, and therefore whether information is wanted in terms of NER or GER expressed as enthalpy or free energy.
"(2) Choose a system boundary. This is facilitated by making a flow sheet depicting inputs and outputs.

"(3) Identify all inputs.
"(4) Assign energy requirement (NER or GER) to all inputs.
"(5) Identify all outputs.
"(6) Establish criteria for partition."

The last step is necessary only when there is more than a single product or service.

Concerning the determination of a system boundary across which to identify and quantify inputs, members of the second IFIAS conference (1975) stated: "Energy analysis often begins with a focus on a product and the process stage by which it is fabricated from material inputs. The system boundary can be defined so as to contain this process stage and no other, and energy analysis would then calculate how much energy is required to carry out this single step. But the system could be defined to account for the energy used to prepare material inputs that are in themselves fabricated in prior stages. Another choice of system boundary would include the final process stage and the processes that generate the inputs to the final stage. A further regression would have the boundary enclose all of these activities plus those that produce the inputs to the stages that yield the fabricated first-stage inputs. This regression can be continued upstream until the system boundary encloses stages that employ only raw materials. Downstream, the boundary may or may not include discard or recycling."

Process analysis has its imperfections, however. Leach (1976) commented concerning process analysis that ". . . unless it is complete it can be misleading. Errors can arise either by concentrating on only a few firms or factories (which may be most untypical) in the skein or by cutting off the 'trace-back' at a premature stage (probably due to exhaustion or old age)." Bullard (1975) also pointed out the danger of double-counting such as coal and electric power produced from coal, and the fact that a growing economy would indicate greater energy requirements than a static one.

Numerous workers have used process analysis in agricultural energy analyses. Barry Commoner *et al.* (1975), in determining the cost of energy used in producing 14 field crops under 29 production situations, utilized detailed crop and energy budget tables. Smerdon *et al.* (1975) prepared fuel use budgets (direct energy consumption only) for numerous Florida production systems. Roller *et al.* (1975) analyzed the energy requirements for five livestock production systems including indirect energy components, the energy required for feeds and the energy required for maintaining breeding stock, and determined energy intensities ranging from 15.0 to 19.4 $MJ \cdot kg^{-1}$ (chickens were most energy efficient and sheep were least). Singh and Chancellor (1975) compared six Indian agricultural production systems and found that, as mechanization level increased, energy intensity increased and labor inputs and production costs per unit of crop output decreased. Avlani and Chancellor (1977) analyzed energy consumption in wheat production and utilization and found that nonirrigated wheat required less energy per unit of production than that which had been irrigated.

Net Energy Analysis (NEA) and Agriculture

Net energy is the difference between the gross energy output from a primary energy source or the energy saved by a conservation practice and that energy required (sometimes termed feedback) to obtain or produce it. Net energy is, then, the output in excess of the feedback. It takes energy to produce or conserve energy. For an operation whose objective is to obtain or conserve energy, and whose outputs and inputs or feedback are of the same quality, the operation must obtain or conserve more energy than is invested as feedback for it to be feasible.

Cottrell (1955) was an early identifier of the net energy concept: ". . . surplus energy . . . is the energy available to man in excess of that expended to make energy available." His two examples: "A stroller eating blackberries growing wild along the road expends in the operations necessary to secure the berries only a small part of the energy he receives from them. He has gained surplus energy. On the other hand, a man who runs down a jack rabbit in an 80-acre field will probably expend more energy than he will gain from the operation."

Howard Odum has been widely identified with the concept of net energy (1971–1977). Odum and Odum (1976) defined net energy as the amount of energy that remains for consumer use after the (energy) costs of finding, producing, upgrading and delivering the energy have been paid. Figure 4.3 shows net energy as the gross output Y minus the feedback F. Net energy equals Y minus F.

Odum and Odum (1976) defined energy yield ratio (EYR) as the output Y from an energy source divided by the feedback energy F required to produce it, both in fossil fuel equivalents. The larger the energy yield ratio the better the source. Odum *et al.* (1976) used this concept to evaluate numerous primary energy surces, i.e., Near East oil obtained by exchange (1975) has an energy yield ratio of 5.7, Gulf of Mexico oil is 6.0, Alaska oil is 6.3, Western coal with 1000 mile transport is 10.6 and hydroelectric power is 19.

Other energy analysts view energy production sectors in a somewhat different but basically equivalent manner. Bullard's (1975) depiction treats inputs as the sum (GER) of feedback and that extracted from the earth and treats outputs as sequestering the sum of the inputs. For net

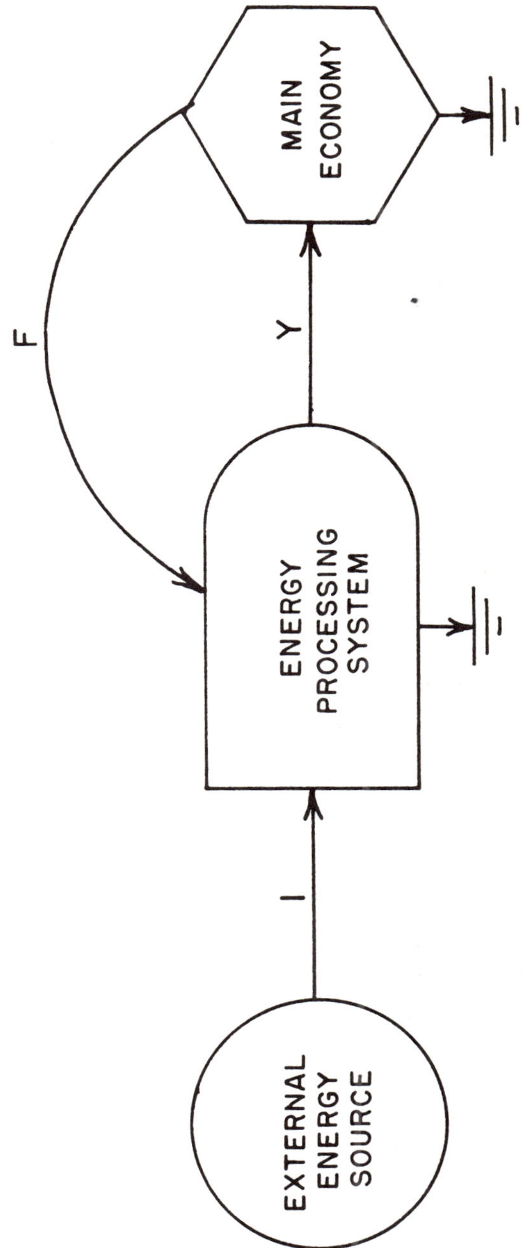

FIG. 4.3. ENERGY PRODUCTION AND PROCESSING, SHOWING NET ENERGY, Y – F

energy to exist, the energy of the output must exceed the feedback energy. Herendeen and Bullard (1974) and Bullard *et al.* (1976) gave the primary energy sequestered ($MJ \cdot MJ^{-1}$) in coal mining and in crude petroleum and natural gas, from which energy yield ratios can be calculated (1 ÷ ("primary" − 1)) (Table 4.10). Slesser (1976) calculated GERs for several synthetic fuel energy technologies and presented values indicating several to be infeasible (GER ≥ 2 or EYR ≥ 1).

Leach (1975) has criticized NEA for several reasons:

(1) Its inability to measure the efficiency of source utilization (it instead compares the usefulness of energy resources).
(2) The difficulty of drawing boundaries across which to accurately measure feedback energy (gasoline stations, new towns for oil shale workers, a Department of Energy, etc., should be charged to the energy sector(s) but they would normally not and are instead viewed as other goods within the GNP).
(3) The meaninglessness of energy yield ratio since, in principle, the boundary can be drawn sufficiently large so that no feedback is required (all inputs self-manufactured) in which case energy yield ratio becomes infinity.
(4) Its difficulty, if not inability, to assign energy yield ratios where there are joint products (i.e., phosphate and uranium from phosphate mining).

Odum *et al.* (1976) and Odum (1977A,B) have identified some energy sources as being secondary as opposed to primary; such sources can supply additional energy but only at the expense of an energy subsidy from a primary energy source. Agriculture is an example of a secondary energy source, solar. Odum and Odum (1976) defined energy investment ratio (EIR) as the ratio of the feedback energy F to that energy supplied by the secondary source I, both measured in FFEs, and suggested that for the United States this quantity is about 2.5.

One difficulty is evaluating the energy yield ratio (EYR) of agriculture, as a secondary energy source, is that so little is known about the FFE of agricultural products. If we hypothesize that the FFE of corn is 10, for

TABLE 4.10. ENERGY SEQUESTERED IN PRIMARY ENERGY SOURCES AND CORRESPONDING ENERGY YIELD RATIOS

	1963		1967	
	primary	EYR	primary	EYR
Coal	1.0142	70.4	1.0092	108.7
Crude	1.0403	24.8	1.0604	16.6

Source: Herendeen and Bullard (1974) and Bullard *et al.* (1976).

instance, and assume that the FFE of the inputs enumerated by Pimentel *et al.* (1973) is unity, then for 1970 the energy yield ratio of United States corn production would have been 28.2 rather than 2.82 as found. Three bits of information on FFE of agricultural products, in possible disagreement with each other, are the statement of Odum *et al.* (1976) that high energy agriculture does not yield net energy and their listings of the FFE of supermarket food as 10 (Odum and Odum 1976) and 24 (Odum *et al.* 1976). Many problems exist in the potential evaluation of the FFE of agricultural products, i.e., are FFE equal for identical products produced by two distinctly different systems with widely differing inputs? Until methodology is developed to determine FFE of agricultural products, little can be said about the energy yield ratios of industrialized agricultural systems, even for those products for which it may have importance.

Primitive subsistence agriculture, in order to survive, must demonstrate the following with respect to NEA: Energy yield ratio must be at least one. The yield Y in kilocalories of food produced must, over time, be at least equal to the feedback F in kilocalories of food consumed. As has been shown by Norman (1978), it should be in the order of 15 to assure long-term survival. Odum (1977B) has termed such agricultural systems as primary sources.

Industrialized agriculture has the following characteristics with respect to NEA: Since industrialized agriculture is a secondary rather than a primary energy source, it need not in general demonstrate an energy yield ratio of at least one. One exception exists: If agriculture is to serve as a source of energy in providing biomass for methane generation, wood for electric power generation, etc., in which case EYR must be greater than one.

BIBLIOGRAPHY

AVLANI, P.K. and CHANCELLOR, W.J. 1977. Energy requirements for wheat production and use in California. Trans. ASAE *20* (3) 429–437.

BAYLEY, S.E., ODUM, H.T. and KEMP, W. 1976. Energy evaluation and management alternatives for Florida's east coast. Trans. 41st N. Amer. Wildlife and Nat. Res. Conf., March 21–25. Washington, D.C.

BEARDSWORTH, E. 1975. A perspective on energy accounting. Ppr. presented AIChE Meeting, Nov. 18. Los Angeles.

BLACK, J.N. 1971. Energy relations in crop production—A preliminary survey. Ann. Appl. Biol. *67,* 272–278.

BREIMYER, H.F. 1975. The food-energy balance. Univ. of Mo., Dep. of Agric. Econ. Ppr. *1975–50,* prepared for delivery at Symp. on Population and Food, December. Rome.

BREIMYER, H.F. 1977. Farm Policy: 13 Essays. Iowa State Univ. Press, Ames.

BULLARD, C.W., III. 1975. Energy costs, benefits, and net enegy. CAC Doc. No. *174.* Cntr. for Adv. Computation, Univ. of Ill., Urbana-Champaign.

BULLARD, C., HANNON, B. and HERENDEEN, R. 1975. Energy flow through the United States economy. Chart published by Cntr. for Adv. Computation, Univ. of Ill., Urbana.

BULLARD, C.W., III and HERENDEEN, R.A. 1975. The energy cost of goods and services. Energy Policy *3* (4) 268–278.

BULLARD, C.W., PENNER, P.S. and PILATI, D.A. 1976. Energy Analysis Handbook. CAC Doc. No. *214,* Cntr. for Adv. Computation, Univ. of Ill., Urbana.

CARTER, A.P. 1974. Applications of input-output analysis to energy problems. Science *184,* 325–329.

CERVINKA, V. *et al.* 1975. Methods used in determining energy flows in California agriculture. Trans. ASAE *18* (2) 246–251.

CHAPMAN, P.F. 1974A. Energy costs: A review of methods. Energy Policy *2* (2) 91–103.

CHAPMAN, P.F. 1974B. The energy costs of fuels. Energy Policy *2* (2) 232–243.

COMM. AGRIC. PROD. EFFICIENCY. 1975. Agricultural production efficiency. Natl. Acad. Sci., Washington, D.C.

COMMONER, B., GERTLER, M., KLEPPER, R. and LOCKERETZ, W. 1975. The vulnerability of crop production to energy problems. *CBNS-AE-2.* Cntr. for Biol. of Nat. Syst., Washington Univ., St. Louis.

COTTRELL, F. 1955. Energy and Society. McGraw-Hill, New York.

DEWIT, C.T. 1975A. Substitution of labour and energy in agriculture and options for growth. Neth. J. Agric. Sci. *23* (2) 145–162.

DEWIT, C.T. 1975B. Agriculture's uncertain claim on world energy resources. Span *18* (1) 2–4.

DOERING, O.C., III. 1974. Energy policy issues for agriculture. *In* Increasing Understanding of Public Problems and Policies—1974. Farm Foundation, Chicago.

DOERING, O.C., III and PEART, R.M. 1977. Evaluating alternative energy technologies in agriculture. NSF/RA-770127, Rept. to Natl. Sci. Found., Purdue U. Agric. Exp. Sta., West Lafayette, Indiana.

FLUCK, R.C. 1975. Energy consumption of the Florida vegetable industry. Proc. Fla. State Hort. Soc. *88,* 128–133.

FLUCK, R.C. 1979. Energy productivity; A measure of energy utilisation in agricultural systems. Agric. Syst. *4* (1) 29–37.

FLUCK, R.C. 1979. Net energy analysis of the energy sequestered in agricultural labor. (Manuscript submitted for publication to Energy Policy)

FED. ENERGY ADMIN. (FEA). 1975. Energy conservation—The data base: The potential for energy conservation in nine industries. Fed. Energy Admin., Washington, D.C.

FEA. 1976. Energy and U.S. agriculture: 1974 data base. FEA/D-76/459, Fed. Energy Admin.—U.S. Dep. Agric., Washington, D.C.

GEORGESCU-ROEGEN, N. 1971. The Entropy Law and Economic Process. Harvard Univ. Press, Cambridge, Mass.

GEORGESCU-ROEGEN, N. 1975. Energy and economic myths. Sou. Econ. J. *41* (3) 347–381.

GIFFORD, R.M. 1976. An overview of fuel used for crops and national agricultural systems. Search *7* (10) 412–417.

GILLILAND, M.W. 1975. Energy analysis and public policy. Science *189,* 1051–1056.

HANNON, B., HARRINGTON, C., HOWELL, R.W. and KIRKPATRICK, K. 1976. The dollar, energy and employment costs of protein consumption. CAC Doc. No. *182,* Cntr. for Adv. Computation, Univ. of Ill., Urbana.

HEICHEL, G.H. 1973. Comparative efficiency of energy use in crop production. Bull. *739,* Conn. Agric. Exp. Sta., New Haven.

HEICHEL, G.H. 1976. Agricultural production and energy resources. Amer. Sci. *64* (1) 64–72.

HERENDEEN, R.A. 1973. The energy cost of goods and services, ONRL-NSF-EP-58. Oak Ridge Natl. Lab., Oak Ridge, Tenn.

HERENDEEN, R.A. and BULLARD, C.W., III. 1974. Energy cost of goods and services, 1963 and 1967. CAC Doc. No. *140.* Cntr. for Adv. Computation, Univ. of Ill., Urbana-Champaign.

HERENDEEN, R. and SEBALD, S. 1974. The dollar, energy, and employment impacts of certain consumer options, Vol. 1. CAC Doc. No. *97.* Cntr. for Adv. Computation, Univ. of Ill., Urbana-Champaign.

HILL, K.M. and WALFORD, F.J. 1975. Energy analysis of a power generating system. Energy Policy *3* (4) 306–317.

HILL, L.D. and ERICKSON, S. 1975. Economic restraints on the reallocation of energy for agriculture. *In* Energy, Agriculture and Waste Management. W.J. Jewell (Editor). Proc. 1975 Cornell Agric. Waste Mgt. Conf., Ann Arbor Science Publishers, Ann Arbor, Michigan.

HIRST, E. 1973. Energy use for food in the United States. ORNL-NSF-EP-57, Oak Ridge Natl. Lab., Oak Ridge, Tennessee.

HOFFMAN, K.C. 1975. Some comments on energy accounting. Paper presented at AIChE Meeting, November. Los Angeles.

HUETTNER, D.A. 1976. Net energy analysis: An economic assesment. Science *192,* 101–104.

INT. FED. INST. ADV. STUDY. (IFIAS). 1974. Energy analysis workshop on methodology and conventions. Int. Fed. of Inst. for Adv. Stud. Wrkshp. Rept. No. *6,* Stockholm.

IFIAS. 1975. Workshop on energy analysis and economics. Int. Fed. of Inst. for Adv. Stud. Rprt. No. *9,* Stockholm.

LEACH, G. 1975. Net energy analysis—Is it any use? Energy Policy *3* (4) 332–344.

LEACH, G. 1976. Energy and Food Production. IPC Science and Technology Press, Guildford, Surrey, U.K.

LEACH, G. and SLESSER, M. 1973. Energy equivalents of network inputs to food production processes. Univ. of Strathclyde, Glasgow.

LOCKERETZ, W. 1975. Consumption of agricultural resources in the production of fat cattle and food grains. J. Soil and Water Cons. *30* (6) 268–271.

MAKHIJANI, A. 1975. Energy and Agriculture in the Third World. Ballinger Publishing Co., Cambridge, Mass.

MAKHIJANI, A.B. and LICHTENBERG, A.M. 1972. Energy and well-being. Environment *14* (5) 10–18.

NORMAN, M.J.T. 1978. Energy inputs and outputs of subsistence cropping systems in the tropics. Agro-Ecosystems *4* (3) 355–366.

NORMAN, M.J.T. 1979. Annual Cropping Systems in the Tropics: An Introduction. Univ. of Florida Press, Gainesville. (in press)

ODUM, H.T. 1971. Environment, Power and Society. John Wiley & Sons, New York.

ODUM, H.T. 1973. Energy, ecology and economics. Ambio *2,* 220–227.

ODUM, H.T. 1977A. Letter to editor. Science *196,* 261.

ODUM, H.T. 1977B. Energy analysis, energy quality, and environment. Univ. of Fla., Gainesville. (unpublished)

ODUM, H.T. and ODUM, E.C. 1976. Energy Basis for Man and Nature. McGraw-Hill, New York.

ODUM, H.T. *et al.* 1976. Net energy analysis of alternatives for the United States. Hearings before the Subcomm. on Energy and Power of the Comm. on Interstate and Foreign Commerce, House of Rep., Ninety Fourth Cong., March 25 and 26.

PENN, J.B. and IRWIN, G.D. 1977. Constrained input-output simulations of energy restrictions in the food and fiber system. Agric. Econ. Rept. No. *280.* Econ. Res. Serv.—U.S. Dep. Agric., Washington, D.C.

PERELMAN, M. 1972. Farming with petroleum. Environment *14* (8) 8–13.

PIEROTTI, A., KELLER, A.G. and FRITSCH, A.J. 1977. Energy and food. CSPI Energy Series X, Cntr. for Sci. in the Public Interest, Washington, D.C.

PIMENTEL, D. *et al.* 1973. Food production and the energy crisis. Science *182,* 443–449.

PIMENTAL, D. *et al.* 1974. Workshop on research methodologies for studies of energy, food, man and environment. Phase I. Cornell Univ., Ithaca, New York.

RAPPAPORT, R.A. 1976. Pigs for Ancestors. Yale Univ. Press, New Haven, Conn.

RAWITSCHER, M. and MAYER, J. 1977. Nutritional outputs and energy inputs in seafoods. Science *198,* 261–264.

ROLLER, W.L., KEENER, H.M. and KLINE, R.D. 1975. Energy costs of intensive livestock production. ASAE Paper *75-4042.* Ppr. presented at Ann. Meet. of Amer. Soc. of Agric. Eng., St. Joseph, Mich.

SINGH, G. and CHANCELLOR, W. 1975. Energy inputs and agricultural production under various regimes of mechanization in northern India. Trans. ASAE *18* (2) 252–259.

SLESSER, M. 1973. Energy subsidy as a criterion in food policy planning. J. Sci. Food Agric. *24,* 1193–1207.

SLESSER, M. 1976. NEA reexamined. Energy Policy *4* (2) 175–177.

SLESSER, M. 1977. Letter to editor. Science *196,* 259, 261.

SMERDON, E.T. *et al.* 1975. Fuel use estimates for Florida agricultural production, Inst. Food and Agric. Sci., Univ. of Fla., Gainesville.

SPEDDING, C.R.W. and WALSINGHAM, J.M. 1975. Energy use in agricultural systems. Span *18* (1) 7–9.

STEINHART, J.S. and STEINHART, C.E. 1974. Energy use in the U.S. food system. Science *184,* 307–316.

U.S. BUREAU OF CENSUS. 1976. Statistical Abstract of the United States: 1976. U.S. Bureau of Census, Washington, D.C.

WEBB, M. and PEARCE, D. 1975. The economics of energy analysis. Energy Policy *3* (4) 318–331.

5

Energy and Economics

DOMINANCE OF ECONOMICS IN ENERGY RELATED DECISIONS

Economics serves to allocate resources among consumers. It does so by the mechanisms of supply, demand and resultant price establishment for products and services, including energy in its various forms. Some inputs, however, have no costs to their users, i.e., air for combustion, and some costs may not be attributed to their sources, i.e., some forms of pollution.

Few entrepreneurs will knowingly and deliberately make decisions which will result in an economic loss of any consequence, even though such decisions may result in other desirable results. Few managers will conserve energy if, upon summing costs and savings, it costs money to do so. On the other hand, some tradeoffs of economic benefits for certain other desirable results are accepted by many managers. Individual goals, judgements and the ability to evaluate the potential tradeoffs are extremely important to the decision. Certain economic losses may be accepted for such benefits as the good of the environment, conservation of energy, the benefit of future generations, to obtain a greater market share, to save time, to reduce risks, to obtain greater reliability, or for aesthetic purposes. However, not many managers are willing to sustain significant economic losses in order to conserve energy. Energy has not and is not likely to assume the position of a unit of value such as the monetary unit. Therefore, for decisions to be made which result in conservation of energy, greater efficiency of energy use, or greater energy productivity, there generally must also occur an economic benefit. In general, we may say that energy savings are necessary but not sufficient.

What is a most important factor in affecting energy consumption and savings is the cost of energy. As energy rises in cost, particularly relative to the cost of other substitutable inputs, entrepreneurs tend to use less, either by reducing wastage and using energy more efficiently, or by substituting cheaper inputs for energy.

COST OF ENERGY

The cost of energy, in economic or monetary terms or dollars, depends upon numerous factors: supply and demand, the form of the energy, its quality, concentration, pollution associated with its use, government regulations, where it is located, transportability, inflation, technology, etc. Table 5.1 shows the costs of various forms of energy from 1950 to 1976. Note the trends with time: There is a gradual increase in most energy costs until about 1973, and from then costs have increased rapidly, and electric power costs bottomed about 1970.

Costs of energy will likely continue to increase in the future. Appleby (1976) predicted that, as the end of the fossil-fuel age approaches, prices will increase rapidly. For example, he estimated coal prices will be in the order of $1.00 · GJ^{-1} in 2000 and perhaps as high as $1.50 · GJ^{-1}. It now seems that his estimates will be far surpassed; coal averaged $0.83 · GJ^{-1} in the United States in 1976. Appleby also predicted that the cost of fossil-fuel generated electric power would be $4.00 · GJ^{-1} ($0.014 · KWh^{-1}) in 2000, a cost most would prefer to pay now instead of that we are actually paying. Further, Appleby (1976) predicted that increasing energy costs could triple the costs of numerous industrial goods, e.g., steel, glass and paper, could increase the cost of inorganic chemicals and fertilizer 2½ times, and would cause food prices to "increase dramatically." In retrospect, these predicted product cost increases are unrealistically high.

Energy User News (Anon. 1978) recently compiled various estimates of 1985 energy costs (in constant 1975 dollars). Averages are: Natural gas, $2.49 · GJ^{-1}; residual fuel oil, $2.89 · GJ^{-1}; distillate fuel oil, $3.30 · GJ^{-1}; steam coal, $1.45 · GJ^{-1}; and electric power, $9.11 · GJ^{-1}.

Energy costs have increased dramatically, most doubling or tripling since pre 1973 prices, and are likely to continue to increase until a new and significant energy source is found and developed. Predictions are now being made of serious disruptions in supply and severe price increases in the mid 1980s.

TABLE 5.1. UNITED STATES ENERGY COSTS[1] IN DOLLARS (CURRENT) PER GIGAJOULE[2]

Source	1950	1955	1960	1965	1967	1970	1971	1972	1973	1974	1975	1976
GNP/energy consumption	7.98		10.76	12.23		13.87	14.67	15.43	16.57	18.37	20.22	21.68
Electric power sales	5.04	4.62	4.67	4.43		4.42		4.91	5.17	6.38	7.52	8.03
Gasoline (retail)						2.82		3.00	3.12	4.24	4.57	
Gasoline (dealer cost)	1.12	1.20	1.19	1.14		1.31		1.32	1.45	2.27	2.65	2.87
Domestic crude					0.48			0.55	0.64	1.08		1.35
Imported crude				0.86		0.92			1.27	4.84	4.50	
Natural gas utilities sales	0.44	0.49	0.57	0.59		0.61		0.69	0.75	0.90	1.22	1.52
Bituminous coal			0.17	0.16		0.25		0.30	0.34	0.63	0.74	0.83
Natural gas at well	0.060	0.096	0.129	0.143		0.157	0.168	0.172	0.201	0.282	0.412	0.538

[1] Source: All values derived from U.S. Bur. of Census (1976, 1977).
[2] A gigajoule is 0.948 million Btu and therefore approximates the quantity of energy (1 million Btu) commonly used for price comparisons.

EFFECTS OF INCREASING ENERGY COSTS ON AGRICULTURE

Agriculture competes in the market with other consumers for energy. Farmers' energy prices have increased dramatically. Figure 5.1 diagrams the prices paid by farmers for several forms of energy. Table 5.2 shows the effect of primarily increasing energy costs on the price of fertilizers. The increasing costs must, in the long run, be passed on to consumers, or the products will not continue to be available. Therefore, increasing energy costs result in increasing costs of all agricultural products.

A moderating factor on the cost-increasing effects of energy costs on agricultural products is that energy has been historically only a minor component of total food costs. Table 5.3 shows that the direct energy inputs of fuel and oil accounted for 5.1% of all production expenses in 1940 and gradually declined to 2.9% in 1973. These direct energy input costs are likely now near or slightly above their 1940 percentage of total production expenses. Total energy costs as a percentage of production costs, including *all* direct and indirect energy inputs, are likely in the order of 10–15%. Leach (1976) stated that the energy inputs of the typical United Kingdom farmer were about 10% of his gross output.

Lockeretz *et al.* (1975) determined the energy costs of 1974 corn belt production as 11% of the production costs of corn, 8% of the production costs of wheat and 6% of the production costs of soybeans. Included as energy costs were direct energy inputs and the indirect energy inputs for manufacturing agricultural chemicals including fertilizers.

The Economic Research Service (ERS) presents values periodically for the "bill for marketing farm foods," the increased costs added between the "farm gate" and purchase by the consumer. The bill is broken down into components as labor, packaging, transportation, corporate profits, business taxes, etc. According to ERS (1974) the direct energy component of this bill was only about 1.7% of the total plus a portion of transportation which was 8% of the total. Again, the total direct and indirect energy inputs of the food marketing bill are likely about 10 to 15% of the total. Leach (1976) stated that the total energy input for the United Kingdom food system represented about 10% of the final price.

One result of the increasing importance of energy costs in the total costs of goods and services is that the total costs are becoming more dependent upon energy costs and less dependent upon nonenergy costs. A doubling of energy costs now might lead to a 10% increase in total costs whereas if energy costs were 25% of total costs and were doubled, total costs would be increased by 25%. As energy costs increase, total costs tend to follow, and economic choices and choices based on energy considerations will tend to coincide.

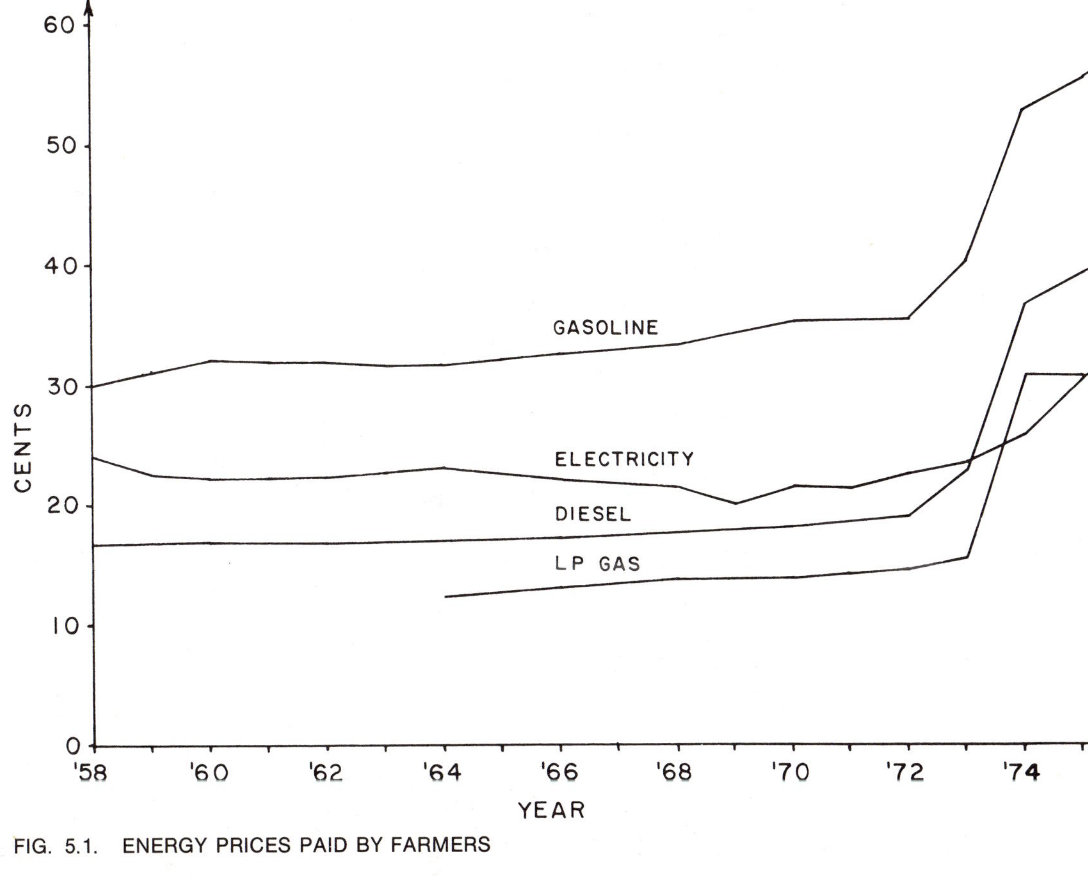

FIG. 5.1. ENERGY PRICES PAID BY FARMERS

Electricity in cents per 10kWh. Others in cents per gallon.

TABLE 5.2. FERTILIZER PRICES

Year	Anhydrous ammonia (S · t^{-1})	Concentrated superphosphate (\$ · t^{-1})	Potash (\$ · t^{-1})
1968	100.7	86.4	54.1
1969	83.3	81.6	52.7
1970	82.7	82.8	56.1
1971	87.4	84.4	64.2
1972	88.2	86.0	64.8
1973	96.6	96.5	67.8
1974	201.7	165.3	89.6
1975	292.1	235.9	112.4
1976	210.5	174.2	105.7
1977	207.2	162.6	106.2

Source: USDA (1977).

Energy price increases are not instantaneously passed through the food system as increased costs of foods. Increased prices for fuels and other consummables are quickly passed on, but increased energy costs for capital items are spread over many years. Steinhart (1974) suggested that at least some of the increased energy costs resulting from the 1973 energy price increases would require some time to work through the system.

Other effects of increased energy prices, including relocation of crop production, were examined by Swanson and Taylor (1977). They found that, in the corn belt, minimization of crop production costs would result in slight shifts in the location of corn acreage, increases in soybean and hay production, and reductions in fertilizer and pesticide use and soil erosion.

Dvoskin and Heady (1976), in a linear programming model with which they examined the effects of several energy constraints on agricultural

TABLE 5.3. FUEL AS A UNITED STATES FARM PRODUCTION INPUT

Year	Total production expenses (million \$)	Petroleum fuel and oil (million \$)	Percentage fuel of total
1940	6,858	350	5.1
1950	19,455	1191	6.1
1960	27,418	1484	5.4
1964	31,715	1567	4.9
1969	42,180	1717	4.1
1970	44,572	1711	3.8
1971	47,603	1722	3.6
1972	52,428	1688	3.2
1973	64,746	1879	2.9

Source: ERS (1974).

production, examined the effects of high energy prices on 1985 agricultural production. A doubling of energy prices from 1974 resulted in only a 5% reduction in energy used in agricultural production (they described this as a low demand elasticity) as compared to no energy constraints. The predominant effects were a 22% reduction in irrigated land and a 14% decrease in purchased nitrogen, which was almost totally compensated for by increases in manure and legume crop utilization. Minor effects were observed on fuel for machinery (1% increase), crop drying (1% decrease), dryland farming (3% increase), and pesticides (2% increase). Perhaps the most significant prediction is that regional farm income would be increased 27% in the South Atlantic (the greatest change) and decrease as much as 18% in the Northwest.

Effects on Texas high plains agriculture were investigated by Adams *et al.* (1976). They determined that increased energy costs would have minimal effect on diesel and natural gas consumption, but that increased costs of nitrogen fertilizer and irrigation water would sharply limit consumption. They further predicted gains in wheat acreage and decreases in grain sorghum and cotton acreage, and decreased profits for farmers with increased energy costs unless agricultural product prices increase.

Connor (1977) also identified several effects which might possibly result from increasing energy costs:

(1) Producers may first reduce energy use.
(2) Enterprise combinations on individual farms may be changed.
(3) Farmers may shift to alternative technologies which result in the substitution of energy inputs or reduction of energy use.
(4) Utilization of new energy sources.
(5) Some farmers may leave farming.

Johnson and Henderson (1977) examined the effects of increased energy costs on irrigation and suggested that, with substantial cost increases, irrigation use could decline, that irrigation might become more efficient, that shifts might occur to crops which require less irrigation, and that further irrigation development might be curtailed.

ENERGY COSTS AND FOOD PRICES

We can predict that increased energy costs will result in increased food costs whenever food results from an industrialized agriculture system. The extent of the effects or the magnitude of the increases are not as easily predicted. In addition to cost increases of foods and other agricultural products we can also expect shifts in production patterns toward less energy intensive crops and livestock. Farmers may shift producing acres from corn to soybeans. The effects of higher energy prices will likely not be the same for all products, on different types of farms, or in different geographic areas.

Steinhart and Steinhart (1974) stated that energy price increases would result in even greater food price increases for societies with industrialized food systems and referenced an article (Slesser 1973) which suggested that a quadrupling of energy prices would result in a sixfold increase in food prices. The interpretation of this statement is open to question. The intention may have been for a literal interpretation, or it may possibly have been that a quadrupling of the cost of the energy sequestered in a product, say from \$0.01 to \$0.04, would result in a price increase of \$0.06. The latter seems much more reasonable now that we have experienced much of the pass-through of increased (quadrupling in the case of imported crude oil) costs to the consumer. Also, the following evidence supports this interpretation.

The ERS (1974) estimated that a doubling of 1971 energy prices would lead to a 6.4% increase in dairy product prices, a 4.8% increase in meat product prices, and a 1.4% increase in grain mill product prices. Much of the effect of energy cost changes occurs within one season's time, but the total effect may not take place for many years. For instance, the energy input for constructing buildings may not be fully passed through until they are replaced or change ownership.

Timmer (1975), in an examination of the agricultural systems of lesser developed countries, developed a simple model by which he predicted that price increases for fertilizer and irrigation energy would ultimately result in an identical increase in food prices.

The only analysis available of the effects of energy cost increases on the entire food system is by Whittlesey and Lee (1976). They determined the effects of energy price changes on 18 fresh and processed foods of Washington origin, including only the direct energy inputs. Inclusion of all indirect energy inputs would have perhaps doubled their increases. They determined that, overall, a doubling of energy prices would result in a 5% increase in food costs, neglecting the possibility of any additional percentage markups. Table 5.4 shows the costs of the direct energy sequestered, from production through home preparation; transportation from retail store to home was not included.

EFFECTS OF ENERGY COSTS ON DIETS

It has been suggested (Steinhart and Steinhart 1974) that increased energy prices would result in reduced energy consumption in the food

TABLE 5.4. COST OF DIRECT ENERGY SEQUESTERED IN FOODS

	Energy cost per unit weight ($ · kg^{-1})	Energy cost per cost of product (%)	Energy per unit weight (MJ · kg^{-1})
Canned peas	0.0311	9.4	0.0120
Frozen peas	0.0631	9.7	0.0207
Fresh potatoes	0.0256	20.8	0.0070
Frozen french fries	0.0639	7.4	0.0278
Dehydrated potatoes	0.1409	7.6	0.0750
Sugar (from sugar beets)	0.2244	12.6	0.0298
Wheat for export	0.0159	8.8	0.0081
Wheat for flour	0.0397	12.3	0.0126
Fresh apples	0.0238	11.0	0.0084
Apple juice	0.0243	8.4	0.0080
Applesauce	0.0289	6.3	0.0127
Dried apples	0.0913	6.3	0.0357
Fluid milk	0.0165	4.5	0.0056
Cheese	0.0653	3.0	0.0252
Cottage cheese	0.0582	3.6	0.0234
Butter	0.0659	3.5	0.0276
Ice cream	≃0.055	4.6	≃0.018
Dried milk	0.0595	4.0	0.0350

Source: Whittlesey and Lee (1976).

system and perhaps a trend away from highly processed and packaged foods. As energy costs increase we may expect to see changes in the diets of those fed by industrialized agriculture. Some of the more energy intensive products (and also those having a more elastic demand) will tend to disappear from the supermarket shelves, or be produced and sold in lesser quantities as luxury foods. Some of the less energy-intensive products will fill that previous demand and become more predominant.

Timmer (1975) stated that, given sufficient time to adjust, food prices will rise to accomodate increased food costs, or else food production will decrease. He also pointed out that in the lesser developed countries the subsistence farmers who do purchase inputs will be affected detrimentally and that the nonfarm poor will find it even more difficult to survive.

Rehkugler (1976), by investigating the minimum food energy required to supply a nutritionally complete diet, has indirectly given us insight into what products may become more predominant in our diets as energy costs increase. He used a linear programming model to select, from among an arbitrary list of foods paralleling menus suggested by a nutritionist, quantities of particular foods which would meet 10 nutritional constraints. Energy coefficients (the energy sequestered in foods) were calculated by multiplying the caloric content of the food by the appropriate energy ratio by Hirst (1973). Rehkugler's conclusions were these:

"(1) Nutrient constraints can be met with minimum energy demand by the consumption of large quantities of relatively few food items.

"(2) If variety of foods in the daily menu and reasonable restrictions on food quantities are introduced as constraints, more foods are introduced in the optimum solution with a resulting higher energy demand and lower energy efficiency.

"(3) As a level of nutrients needed in the diet increased, the energy efficiency decreased and total energy required to supply the diet increased. The higher nutrient quality diet demands more energy input to our food systems.

"(4) Daily menus vary in energy demands when constrained to provide variety of items and acceptable quantities. Therefore, on a weekly cycle it should be possible to emphasize increased consumption during certain days of the week to give a lower overall energy consumption in our food system.

"(5) The differences in female and male nutrient demands causes differences in the minimum energy diet for these two groups. Because females tend to require a higher quality diet to provide the additional iron, they tend to require more energy from the food system to supply one unit of their caloric intake. Their daily total energy demands, however, are less because of a much lower caloric need as compared to males.

"(6) Cereal products and fresh vegetables are energy efficient foods and to minimize energy demands on our food system should be introduced first into our daily diet. Dairy products and meats are the next most important food items to be introduced into the diet to provide the higher level nutrient demands at a minimized energy cost to our food system.

"(7) Finally, note that the above calculations are reached on the basis of the assumed accuracy of the energy coefficients for the various food groups. Further study may show that the relative magnitudes of these values are in error. These results should be considered preliminary until further menus have been individually investigated and then grouped to examine minimum energy over a weekly cycle of menus."

EFFECTS OF CURTAILED ENERGY SUPPLY

The effects of a curtailed energy supply to agriculture, whether the curtailment is due to shortages or governmental allocations, have been examined from various points of view.

At the producer level, Williams and Chancellor (1975) estimated California crop production decreases due to a hypothetical 50% reduction in production fuel (essentially the direct energy inputs) to range from 19 to 37%. The energy cost reduction was $63 million and the production value reduction was $462 million.

Chen and Wensink (1977) modeled optimum Willamette Valley agricultural production under conditions of reduced energy supplies. They found reduced profits, shifted crop production patterns and less digestible protein production.

Dvoskin and Heady (1976) examined the nationwide effects of a hypothetical 10% energy supply reduction. They found that agricultural commodity prices would vary both upward and downward depending upon whether decreased or increased production of the commodity resulted, that farm income and food prices would increase and that agricultural exports would increase.

Jesse (1976) suggested that extreme energy shortages would shift production and consumption away from livestock products and toward grain products. He also suggested that fruit and vegetable production would not be severely affected since those products are more energy efficient than livestock products, require little processing and are desirable to improve diet palatability. He suggested that fresh would increase relative to processed and that root crops would increase.

Penn and Irwin (1977) used a constrained input-output simulation of United States agriculture to which were applied energy restrictions. A simulated reduction of crude oil imports or of refined petroleum products affected agriculture less than other industries. Simulated allocation of all needed refined petroleum to agriculture while other sectors of the economy shared the shortages, however, wreaked havoc by reducing other needed inputs to agricultural production.

FOOD EXPORTS AND OIL IMPORTS—ENERGY BALANCE OF TRADE

One question of current interest is whether it is desirable or possible for the United States to purchase foreign energy with food exports. This strategy has succeeded to a degree, thus far. The values of net annual agricultural exports and of fuel imports are shown in Fig. 5.2. Net agricultural exports have provided for a significant portion of the nation's fuel imports and have risen when fuel imports rose sharply in 1973 and 1974.

Obviously, it required more energy to produced agricultural products for foreign export and domestic needs than is required to meet domestic

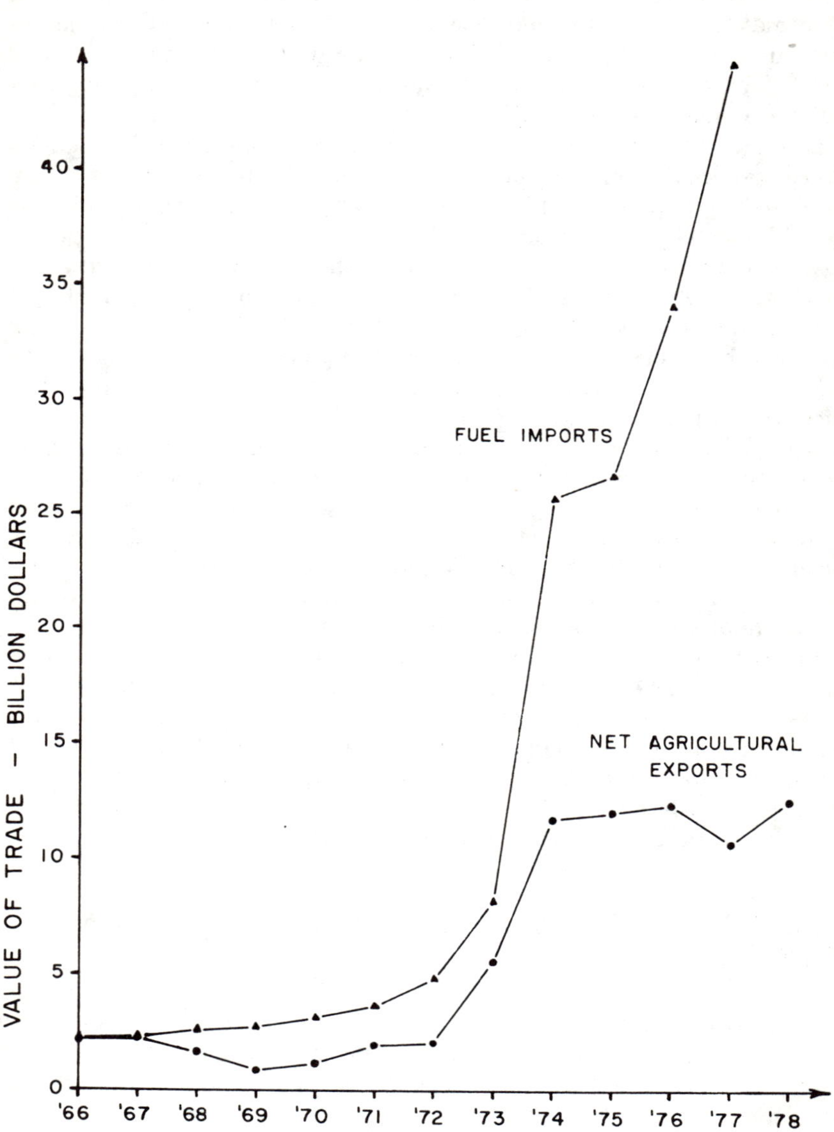

FIG. 5.2 COMPARISON OF THE VALUE OF NET AGRICULTURAL EXPORTS AND FUEL IMPORTS

needs alone. ERS (1974) questioned the advisability of producing agricultural goods for export but presented data showing agricultural products to be among the least energy intensive ($MJ \cdot \$^{-1}$) of all goods produced. It was concluded that agricultural products should be among the last categories of goods to be restricted from export.

Bentley and Long (1976) illustrated the economic feasibility of trading agricultural commodities for petroleum. With oil at $\$14.00 \cdot bbl^{-1}$, wheat at $\$4.00 \cdot bu^{-1}$, and corn at $\$3.00 \cdot bu^{-1}$, they said that only 6.4% of the oil purchased with the proceeds from a bushel of wheat would be required to produce that bushel and 7.9% of the oil purchased would be required to produce corn.

Odum and Odum (1976) pointed out: "If the United States exchanges 50 million tons of wheat for 620 million barrels of oil, the yield to the United States, expressed in fossil fuel equivalents, is 5 calories for 1 calorie sent abroad and 1 calorie used in agricultural work. Industrialized agriculture in the United States can continue temporarily as long as energy inputs can be sustained."

Further, Odum *et al.* (1976) analyzed net energy from trading grain for oil at 1975 prices ($\$3.80 \cdot bu^{-1}$ and $\$12.00 \cdot bbl^{-1}$) and, assuming one kcal of primary energy to be sequestered in 1 kcal of wheat, calculated an energy ratio of 4.4. If the primary energy required to produce grain for export is less than 1 kcal per kcal (Pimentel showed 2.82 kcal return for 1 kcal in production of corn) the energy yield ratio would be greater than 4.4. Also, relative prices will affect the energy yield ratio.

Nelson and Bullard (1975) used input-output analysis to examine the energy content of goods exported and imported. Table 5.5 summarizes the results. It is not known why their values for 1972 net agricultural exports seem in contradiction with other accepted data, indicating net imports rather than net exports. The overall results show that the United States has imported more energy than it has exported, in direct and indirect forms in all traded goods, for each of the three years. Also, the

TABLE 5.5. ENERGY CONTENT AND VALUE OF NET EXPORTS

	GJ			Million Dollars		
	1963	1967	1972	1963	1967	1972
Feed grains	102	124	5	1,682	1,726	77
All agric. products	147	152	– 79	2,068	2,118	– 1,769
All agric. and food products	131	160	– 188	2,376	2,937	– 3,080
Crude petro. and gas	– 2,589	– 2,397	– 5,243	– 1,329	– 1,143	– 3,538
All trade	– 3,852	– 4,085	– 10,836	10,980	13,269	2,016

Source: Nelson and Bullard (1975).

value of all exports has been greater than that of all imports for each of the three years.

Herendeem and Bullard (1975), in a subsequent publication, stated that the United States is in approximate energy balance (at about 6% of domestic energy production) for imported and exported nonenergy goods. However, imported energy in fuels has increased sharply from 13% of domestic production in 1963 to 26% of domestic production in 1973 and near 50% of current total consumption.

BIBLIOGRAPHY

ADAMS, B.M., LACEWELL, R.D. and CONDRA, G.D. 1976. Economic effect on agricultural production of alternative energy input prices: Texas high plains. Tech. Rep. No. *73.* Texas Water Resources Inst., Texas A & M Univ., College Station.

ANON. 1978. Energy User News. *3* (9) 4.

APPLEBY, A.J. 1976. Energy costs and society. Energy Policy *4* (2) 87–97.

BENTLEY, O.G. and LONG, R.W. 1976. A national program of agricultural energy research and development. Natl. Assoc. State Land Grant Colleges and the U.S. Dep. Agric., Washington, D.C.

CHEN, K.L. and WENSINK, R.B. 1977. Optimizing Oregon agricultural production systems with limited energy supplies. Trans. ASAE *20* (5) 986–991.

CONNOR, L.J. 1977. Agricultural policy implications of changing energy prices and supplies. *In* Energy and Agriculture. William Lockeretz (Editor). Academic Press, New York.

DVOSKIN, D. and HEADY, E.O. 1976. U.S. agricultural production under limited energy supplies, high energy prices, and expanding energy exports. CARD Rept. *69.* Cntr. for Agric. and Rural Dev., Iowa State Univ., Ames.

ECONOMIC RES. SERV. (ERS). 1974. The U.S. food and fiber sector: Energy use and outlook. Economic Res. Serv.—U.S. Dep. Agric., Washington, D.C.

HERENDEEM, R.A. and BULLARD, C.W., III. 1975. U.S. energy balance of trade, 1963–1973. CAC Tech. Memo No. *60.* Cntr. for Adv. Computation, Univ. of Ill., Urbana.

HIRST, E. 1973. Energy use for food in the United States. ORNL-NSF-EP-57. Oak Ridge Natl. Lab., Oak Ridge, Tenn.

JESSE, E.V. 1976. Energy inputs in U.S. fruit and vegetable production: Relative costs and efficiencies. Ppr. presented at IX Int. Cong. of NOR-COFEL, November 3–5. Dijon, France.

JOHNSON, B.B. and HENDERSON, P.A. 1977. Energy price levels and the economics of irrigation. Dep. Agric. Econ. Rept. No. *79A,* Univ. of Nebraska, Lincoln.

LEACH, G. 1976. Energy and Food Production. IPC Science and Technology Press, Guildford, Surrey, U.K.

LOCKERETZ, W. *et al.* 1975. Energy in corn belt crop production. CBNS-AE-5, Cntr. for the Biol. of Nat. Sys., Washington Univ., St. Louis.

NELSON, M.W. and BULLARD, C.W., III. 1975. Energy content of traded goods: 1963, 1967, 1972. CAC Tech. Memo. No. *48,* Cntr. for Adv. Computation, Univ. of Ill., Urbana.

ODUM, H.T. and ODUM, E.C. 1976. Energy Basis for Man and Nature. McGraw-Hill, New York.

ODUM, H.T. *et al.* 1976. Net energy analysis of alternatives for the United States. Congressional Record, March 25 and 26. U.S. Govnt. Printing Off., Washington, D.C.

PENN, J.B. and IRWIN, G.D. 1977. Constrained input-output simulations of energy restrictions in the food and fiber system. Agric. Econ. Rept. No. *280,* Economic Res. Serv.—U.S. Dep. Agric., Washington, D.C.

REHKUGLER, G.E. 1976. Energy efficiency in supplying recommended dietary allowance. ASAE Paper *76-6521* presented at Winter Meet. of Amer. Soc. Agric. Eng., St. Joseph, Michigan.

SLESSER, M. 1973. *Cited by* J.S. Steinhart and C.E. Steinhart. Energy use in the U.S. food system. Science *184,* 307–316.

STEINHART, J.S. 1974. Energy policy alternatives and food costs. *In* Increasing Understanding of Public Problems and Policies. Farm Foundation, Chicago.

STEINHART, J.S. and STEINHART, C.E. 1974. Energy use in the U.S. food system. Science *184,* 307–316.

SWANSON, E.R. and TAYLOR, G.R. 1977. Potential impact of increased energy costs on the location of crop production in the corn belt. J. Soil and Water Cons. *32* (3) 126–129.

TIMMER, C.P. 1975. Interaction of energy and food prices in less developed countries. Amer. J. Agric. Econ. *57* (2) 219–224.

U.S. BUR. OF CENSUS, 1976. Statistical Abstract of the United States. U.S. Govn't. Print. Off., Washington, D.C.

U.S. BUR. OF CENSUS. 1977. Statistical Abstract of the United States. U.S. Govn't. Print. Off., Washington, D.C.

U.S. DEP. AGRIC. (USDA). 1977. 1977 Handbook of Agricultural Charts. Agric. Handbk. No. *524,* U.S. Dep. Agric., Washington, D.C.

WHITTLESEY, N.K. and LEE, C. 1976. Impacts of energy price changes on food costs. Bull. *822,* Coll. of Agric. Res. Cntr., Washington State Univ., Pullman.

WILLIAMS, D.W. and CHANCELLOR, W.J. 1975. Irrigated agricultural production response to constraints in energy-related inputs. Trans. ASAE *18* (3) 459–466.

6

Energy Requirements for Agricultural Inputs

PLANT NUTRIENTS: FERTILIZERS AND MANURES

Chemical fertilizers are generally credited with about ½ to ⅔ of the production of industrialized agriculture (Davis and Blouin 1976; Schneeberger and Breimyer 1974). In addition, use of fertilizers increases the partial energy productivities of other inputs by increasing production per unit area. The same plowing, planting, cultivating, etc., are required regardless of fertilizer application and yield. Pratt (1966) pointed out the economic incentive in using fertilizer. The primary macronutrients are nitrogen, phosphorus and potash, the percentage analysis shown on fertilizer labels and bags. The first number is the percentage of N, the second the percentage of P_2O_5 and the third the precentage of K_2O. Removal of crops from land removes these macronutrients from the soil which must ultimately be replenished for continued production. Pratt (1966) indicated that 2240 $kg \cdot ha^{-1}$ of wheat is equivalent to 18 kg of nitrogen, 3.6 kg of phosphorus and 4.0 kg of potassium.

Chemical fertilizers are energy intensive. They require considerable quantities of fossil fuels to produce (Table 6.1). In particular, the production of ammonia for nitrogen fertilizers is heavily dependent upon natural gas. In addition, due to the large quantities of chemical fertilizers which are applied, they consume more indirect energy than any other agricultural input. According to the Federal Energy Administration (FEA 1976), the United States consumed 0.655×10^{12} MJ of energy for fertilizers used in 1974. Faidley (1977) estimated 0.75×10^{12} MJ for

TABLE 6.1. ENERGY REQUIREMENTS FOR CHEMICAL FERTILIZERS IN MEGAJOULES PER KILOGRAM OF N, P_2O_5, or K_2O

Nutrient and Form	Energy ($MJ \cdot kg^{-1}$)	Source
Nitrogen		
Ammonia:		
Steam/air/methane 1966	43.5	Leach and Slesser (1973)
Steam/air/methane 1970	35.0	Leach and Slesser (1973)
Steam/air/methane 1970	32.8	Leach and Slesser (1973)
Steam/air/methane 1970	35.0	Leach and Slesser (1973)
Braun process	32.7	Leach and Slesser (1973)
Atlantic Richfield plant, Iowa	37.9	Leach and Slesser (1973)
Steam reforming	48.6	Blouin and Davis (1975)
Naptha reforming	48.1	Blouin and Davis (1975)
Heavy oil partial oxidation	48.8	Blouin and Davis (1975)
Coal gassification	67.7	Blouin and Davis (1975)
Unspecified	47.6	ERS (1974)
Unspecified	47.7	Miles (1975)
Unspecified	56.3	Commoner *et al.* (1974)
Unspecified	57.3	Hoeft and Siemens (1975)
Unspecified	57.0	Sherff (1975)
Urea:		
Stamicarbon	74.7	Leach and Slesser (1973)
Chemico	74.5	Leach and Slesser (1973)
Montecatini	76.0	Leach and Slesser (1973)
Metrui	74.2	Leach and Slesser (1973)
Snam-Progetti	70.9	Leach and Slesser (1973)
Unspecified	64.7	Blouin and Davis (1975)
Solid urea	50.1	ERS (1974)
Solution urea	57.3	ERS (1974)
Urea prills	65.9	Miles (1975)
Urea solution	71.9	Hoeft and Siemens (1975)
Solid urea	77.7	Hoeft and Siemens (1975)
Solution	67.4	Sherff (1975)
Prilled	72.1	Sherff (1975)
Ammonium nitrate:		
Nitric acid	62.0	Leach and Slesser (1973)
I.C.I Nitram-process	61.8	Leach and Slesser (1973)
Most modern process	59.1	Blouin and Davis (1975)
Industry average	71.1	Blouin and Davis (1975)
Unspecified	51.5	ERS (1974)
Unspecified	59.1	Miles (1975)
Unspecified	76.5	Commoner *et al.* (1974)
Solution	65.3	Hoeft and Siemens (1975)
Solid	76.7	Hoeft and Siemens (1975)
Solution	62.8	Sherff (1975)
Prilled	72.1	Sherff (1975)
Ammonium sulfate:	66.2	Leach and Slesser (1973)
Unspecified	125.6	Dekkers *et al.* (1974)
Unspecified	70.0	Stansfield (1975)
Unspecified	80.0	Schuffelen (1975)

TABLE 6.1. *(Continued)*

Nutrient and Form	Energy (MJ · kg^{-1})	Source
Unspecified	77.5	Pimentel *et al.* (1973)
Unspecified	63.3	Cervinka *et al.* (1975)
Unspecified	83.1	Biswas and Biswas (1975)
Unspecified	80.0	Faidley (1977)
Phosophorus		
Triple superphosphate using Frasch-mined sulfur	9.7	Blouin and Davis (1975)
Triple superphosphate using recovered sulfur	3.6	Blouin and Davis (1975)
Triple superphosphate	11.0	Miles (1975)
Triple superphosphate	11.9	Commoner *et al.* (1974)
Triple superphosphate	22.1	Sherff (1975)
Normal and enriched superphosphate	0.9	ERS (1974)
Normal superphosphate	17.4	Sherff (1975)
Concentrated superphosphate	2.9	ERS (1974)
Phosphate	4.7	Miles (1975)
Unspecified	0.8	Cervinka and Chancellor (1975)
Unspecified	22.1	Coble and Lepori (1974)
Unspecified	13.4	Biswas and Biswas (1975)
Unspecified	33.8	Dekkers *et al.* (1974)
Unspecified	14.0	Pimentel *et al.* (1973)
Unspecified	14.0	Stansfield (1975)
Unspecified	12.0	Schuffelen (1975)
Unspecified	14.0	Faidley (1977)
Potash		
By shaft mining	3.7	Blouin and Davis (1975)
By solution mining	14.4	Blouin and Davis (1975)
Potassium chloride	2.8	ERS (1974)
Potassium chloride	7.0	Sherff (1975)
Potassium sulfate	1.6	ERS (1974)
Muriate of potash	4.5	Miles (1975)
Unspecified	4.9	Commoner *et al.* (1974)
Unspecified	9.2	Biswas and Biswas (1975)
Unspecified	4.6	Dekkers *et al.* (1974)
Unspecified	9.7	Pimentel *et al.* (1973)
Unspecified	8.0	Stansfield (1975)
Unspecified	8.0	Schuffelen (1975)
Unspecified	9.0	Faidley (1977)
Combinations*		
Diammonium phosphate	45.1	Leach and Slesser (1973)
Diammonium phosphate using Frasch-mined sulfur	34.5	Blouin and Davis (1975)
Diammonium phosphate using recovered sulfur	14.6	Blouin and Davis (1975)
Diammonium phosphate	73.9	Miles (1975)
Diammonium phosphate	25.9	Commoner *et al.* (1974)
Diammonium phosphate	70.9	Sherff (1975)

* Energy per unit weight of sum of nutrients.

North America in 1972–1973 and 3.381×10^{12} MJ worldwide for the same year.

The response of a crop to fertilizer depends upon the amount applied. As the level of fertilization increases, yield increases typically become smaller and perhaps even peak and then diminish. Pratt has supplied an example in Fig. 6.1. Even at the highest application levels, it may be economic to apply additional nitrogen (at an application rate of 400 $kg \cdot ha^{-1}$, an additional 45 $kg \cdot ha^{-1}$ costing \$4.00 returns an additional 125 $kg \cdot ha^{-1}$ of corn worth \$6.00). Recommended application rates of nitrogen on nonirrigated Corn Belt corn are in the order of 225 $kg \cdot ha^{-1}$. Pimentel (1974) cited data from Couston showing the response to nitrogen fertilizer inputs of wheat in India and rice in the Philippines. Both are shown in Fig. 6.2.

Several have been concerned with the crop energy return to the application of fertilizer. Hoeft and Siemens (1975) indicated net energy increases up to the point of near maximum yield. Stansfield (1975) pointed out that in the United States between 1950 and 1970, fertilizer energy input increased from 36 to 250 $MJ \cdot ha^{-1}$ but that the energy content of crops increased from 910 to 1950 $MJ \cdot ha^{-1}$, "surely a worthwhile way of capturing solar energy more effectively." Schuffelen (1975) in his analysis of this question concluded that there is always an energy gain so long as fertilizers are applied mainly for primary production of arable land crops.

According to Blouin and Davis (1976), about 95% of all nitrogen fertilizer in the United States is produced from synthetic anhydrous ammonia by the Haber process. About 70% of the 16 MMT (million metric tons) of ammonia produced annually is used in agriculture. From ammonia, urea and ammonium nitrate are produced. Davis and Blouin (1976) showed the utilization of the various forms of fertilizer nutrients and also detailed energy consumption for product transportation energy (about 6% of production energy), storage and transfer energy (about 5% of production) and field application energy (about 2% of production).

Manures are a partial alternative to the use of chemical fertilizers. Coble and LePori (1974) stated that in 1973 most of the manure produced in the high plains area of Texas was used on crops, plus some of the surplus from previous years. They estimated the energy required for collecting, transporting 16.1 km and spreading 22.4 $t \cdot ha^{-1}$ as 0.14 $MJ \cdot kg^{-1}$ of manure or about 4 $MJ \cdot kg^{-1}$ of fertilizer nutrient. Obviously this can be a less energy intensive nutrient input than chemical fertilizers. However, manures cannot totally substitute for fertilizers as the quantities available are insufficient. Coble and LePori (1974) indicated manures could replace only 15% of the energy consumed in manufacturing anhydrous ammonia for Texas.

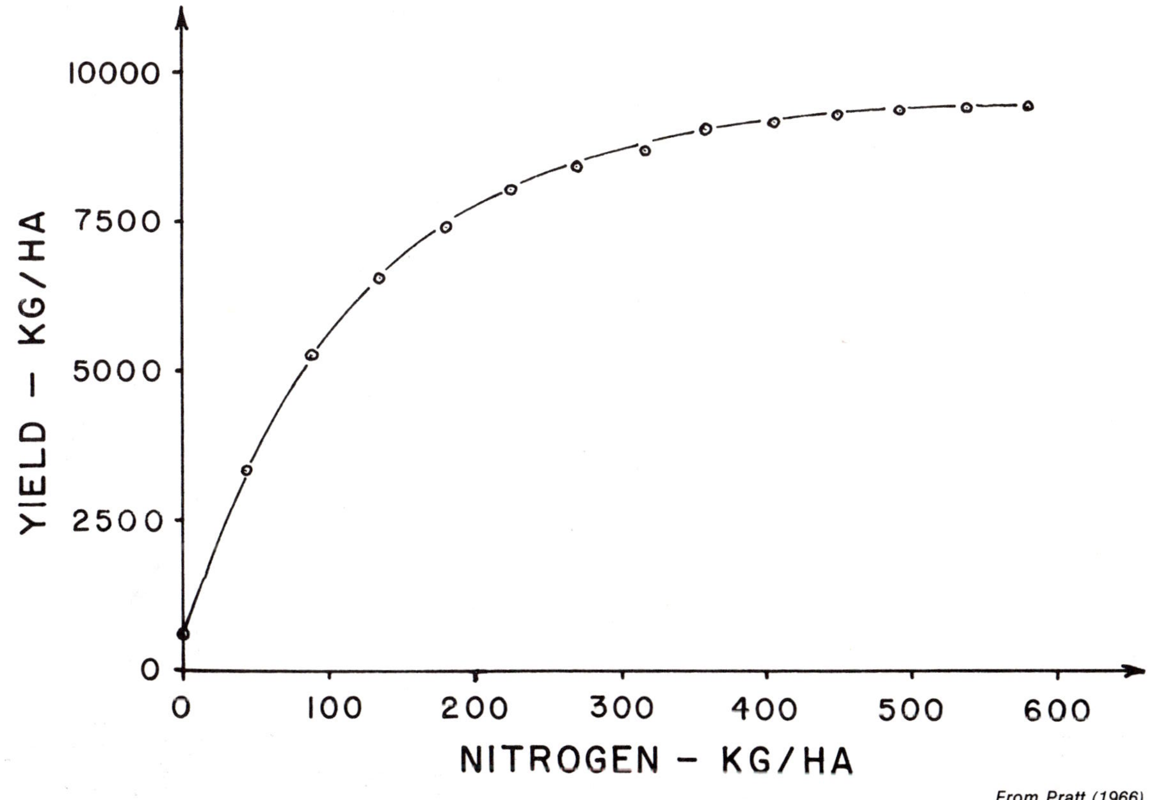

From Pratt (1966)

FIG. 6.1. EFFECT OF NITROGEN FERTILIZER ON CORN YIELD

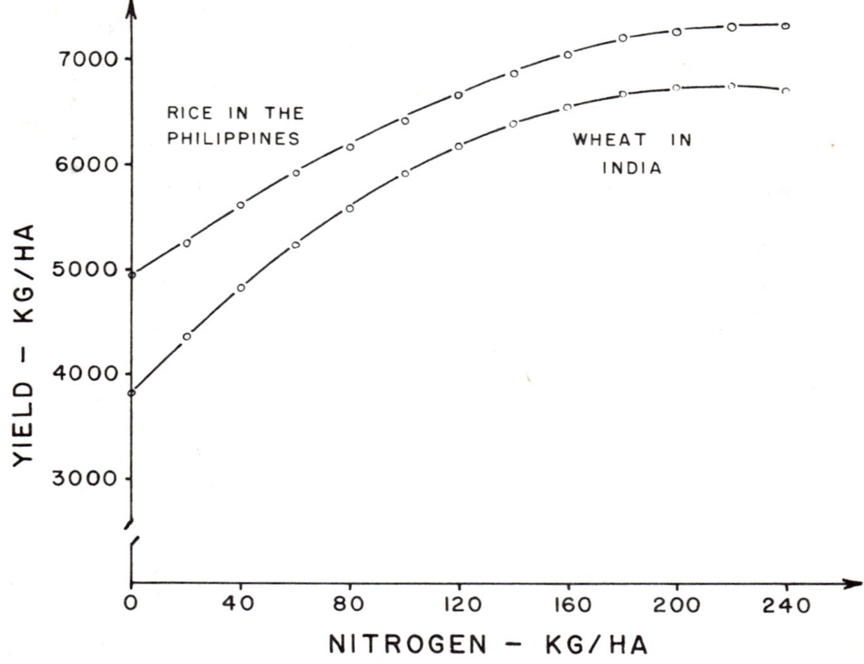

FIG. 6.2. EFFECT OF NITROGEN FERTILIZER ON RICE AND WHEAT YIELDS

Pimentel *et al.* (1973) also extolled the benefits of using manure. With an application rate of 22.4 $t \cdot ha^{-1}$ and a 0.8 to 1.6 km haul they calculated a 73% reduction in energy required to apply the same nutrients as compared to chemical fertilizers.

PESTICIDES

Pesticides include herbicides, insecticides, fungicides, rodenticides and others. The Economic Research Service (ERS 1974) data show 1971 United States farm use of pesticides as being 103 million kg of herbicides, 77 million kg of insecticides, 19 million kg of fungicides, and 25 million kg of other pesticides. In addition, 51 million kg of sulfur and 101 million kg of petroleum were used as or in pesticide preparation. Herbicide use has increased sharply in recent years as an alternative to mechanical cultivation. ERS data (1974) for energy consumed by the pesticide industry, 10.9×10^9 MJ for 1971, apparently does not include the energy requirements for feedstock materials. FEA (1976) reported 100.5×10^9

MJ as the 1974 pesticide energy consumption for the United States. Combining the latter value with estimated 1974 United States pesticide consumption of 272 million kg, yields an average energy requirement of approximately 370 MJ · kg^{-1}. Estimates of the 1972–1973 worldwide energy consumption for pesticides were given by Faidley (1976) at 162.7×10^9 MJ and North American pesticide energy at 93.6×10^9 MJ.

Others have also estimated the energy sequestered in pesticides. Leach and Slesser (1973) used process analysis to determine the quantities 101.5 MJ · kg^{-1} for DDT and 94.2 MJ · kg^{-1} for 2, 4-D. These values were used and generalized by Pimentel *et al.* (1973), Nalewaja (1974), Faidley (1977) and others. Jones (1975) also used this value as well as typical application rates and technology to determine the total energy required for pesticide application at 262 MJ · ha^{-1}. Commoner *et al.* (1974) calculated a value for DDT somewhat higher than Leach and Slesser's: 157 MJ · kg^{-1}. Green and McCulloch (1976) calculated the direct energy requirements for several herbicides: MCPA, 130 MJ · kg^{-1}; Diuron, 270 MJ · kg^{-1}; Atrazine, 190 MJ · kg^{-1}; Triflurolin, 150 MJ · kg^{-1}; and Paraquat, 460 MJ · kg^{-1}. Green (1976) also published a more extensive listing (Table 6.2). Edminster (1975) said that 326 MJ were required to manufacture and apply 16.3 kg of atrazine to 1 ha of land. He further quoted another source as calculating that 507 MJ · ha^{-1} would supply herbicides at the rate of 4.5 kg · ha^{-1}.

TABLE 6.2. ENERGY REQUIREMENTS FOR PESTICIDES

Chemical	MJ · kg^{-1}
MCPA	127.7
Diuron	274.5
Atrazine	188.3
Trifluralin	150.9
Paraquat	459.4
2,4-D	87
2,4,5-T	135
Chloramben	170
Dinoseb	80
Propanil	220
Propachlor	290
Dicamba	295
Glyphosate	454
Diquat	400
Ferbam	81
Maneb	99
Captan	115
Methyl parathion	160
Toxaphene	58
Carbofuran	454
Carabaryl	153
Phorate	172

Source: Green (1978).

Published data thus yields energy requirements for various pesticides within a range of 58 to 460 $MJ \cdot kg^{-1}$. Signficant additional quantities of energy may be required in the mixing of some pesticides with an oil in preparation for application. For instance, a pesticide requiring 200 $MJ \cdot kg^{-1}$ for the active ingredient and mixed one part pesticide to four parts oil base at 45 $MJ \cdot kg^{-1}$ will sequester 380 $MJ \cdot kg^{-1}$ of active ingredient. Also, additional energy is required for packaging, transportation, distribution, etc. (Pimentel 1976).

MACHINERY AND FUEL

Energy required for the use of agricultural machinery can be categorized into that required for providing the machinery, including maintenance and repairs, and that for fuel. The energy sequestered in manufacturing agricultural machinery should be attributed to or amortized over perhaps numerous enterprises over the several years of expected life of the machinery. Major repairs, such as an engine overhaul, should be treated similarly. Maintenance should be charged to the enterprises to which the machinery is assigned during the period the maintenance applies. Fuels should be charged to their immediate use.

Energy requirements for manufacturing agricultural machinery can be estimated by several methods. Estimations can be based on the cost of the machine by multiplying by a transformation such as the energy consumption to GNP ratio of 49.89 $MJ \cdot \$^{-1}$; costs of machinery are readily obtainable from such publications as the *National Farm Tractor and Implement Blue Book* (Anon. 1978). More accurate values of energy intensity of farm machinery are available from Bullard *et al.* (1976); their procedures also allow for distribution of purchase costs and energy requirements among manufacturing, transportation, wholesaling, retailing and for adjusting for inflation.

Process analysis is another method for estimating the energy requirements for manufacturing machinery. Several (Faidley 1977; Pimentel *et al.* 1973) have used the analysis of Berry *et al.* (1973) for automobile manufacturing as an estimate of the energy sequestered in agricultural machinery. Berry *et al.* (1973) found the energy intensity of automobile manufacturing to be 86.8 $MJ \cdot kg^{-1}$. Weights of farm equipment necessary for an estimate of energy requirements are also published in the *Blue Book* (Anon. 1978). It is likely, however, that an average piece of farm equipment is not as energy intensive as an automobile, although tractors and other self-propelled equipment may approach this energy intensity.

Doering *et al.* (1977) estimated agricultural machinery energy requirements on a value added basis (equivalent to IFIAS's NER) and arrived at the values of energy intensity shown in Table 6.3. Their analysis basically excludes the energy sequestered in the metal. Addition of the energy sequestered in the metal would add in the order of 30–60 $MJ \cdot kg^{-1}$ to the above values, and perhaps also change the relative values for the implements listed.

Total annual worldwide energy consumption for manufacturing agricultural machinery was estimated by Faidley (1977) at 1.304×10^{12} MJ with North American consumption at 0.429×10^{12} MJ. Faidley estimated annual worldwide fuel consumption at 2.582×10^{12} MJ and North American fuel consumption at 0.870×10^{12} MJ. FEA (1976) estimated United States 1974 fuel consumption for machinery operations associated with crops from preplant through harvesting to be 0.412×10^{12} MJ. Some additional fuel is consumed by field machinery in operations such as transportation and livestock feeding and waste handling. Estimates of the total United States fuel consumed in agricultural production annually center around 0.90×10^{12} MJ and include a substantial portion for farm vehicles.

Fuel consumption may be estimated using the procedure in the *Agricultural Engineers Yearbook* (ASAE 1978) or, for a specific tractor, the Nebraska Tractor Tests data. The Nebraska Tractor Tests are published in *Agricultural Engineers Yearbook* and the annual January 31 "Red Book" issue of *Implement & Tractor* (Anon. 1979). A 100 PTO Hp diesel tractor might consume 16.6 $L \cdot h^{-1}$ of diesel fuel as calculated by one ASAE formula, or 21.4 $L \cdot h^{-1}$ by another formula when operated specifically at half load. The average of 10 test reports on diesel tractors in the 100 PTO HP range at 50% load was 17.3 $L \cdot h^{-1}$ during which they produced an average of 48.9 drawbar horsepower.

TABLE 6.3. ENERGY REQUIREMENTS FOR MANUFACTURING FARM MACHINERY, EXCLUDING ENERGY SEQUESTERED IN METALS

Implement	$MJ \cdot kg^{-1}$
Tractor	27.63
Combine	21.65
Plow	12.78
Disc	9.96
Applicator	10.20
Planter	16.90
Rotary Hoe	12.38

Source: Doering *et al.* (1977).

The actual energy sequestered in a petroleum product such as diesel fuel is more than the energy content of the fuel. Bullard *et al.* (1976) gave a primary value $[(Y + F) \div Y]$ of 1.2227 for petroleum refinery products. The indicated energy, therefore, sequestered in diesel fuel is $1.2227 \times 38.7\ MJ \cdot L^{-1}$ or $47.3\ MJ \cdot L^{-1}$.

Engine oil and filter costs are estimated by ASAE (1978) as 15% of total fuel costs. Since unit oil costs are in the order of four times unit fuel costs, oil and filter energy consumption are estimated to be in the order of 4% of engine fuel energy.

Another energy consuming input for agricultural machinery is repairs. Doering *et al.* (1977) estimated repair energy to be about 33% of manufacturing energy over a 10 year period. Although several have taken repairs to be only 6% of initial cost for farm machinery, based on automobile data[1] (Berry *et al.* 1973), higher and more realistic amounts are indicated from formulas given by ASAE (1978). Total accumulated repair costs depend heavily upon use; over the total life of the machine repairs can easily amount to as much as the initial cost. For instance, the formula for a diesel tractor's repair and maintenance costs

$$C_R = 0.120\ X^{2.033} C_I$$

Where: $X =$ accumulated thousands of hours of use and $C_I =$ initial cost, which gives, for $X = 10{,}000$ hours, $C_R = 1.29\ C_I$. Total repair costs during a typical 10,000 hour diesel tractor lifetime are thus expected to be 129% of the initial cost. Further, repairs are probably about as energy intensive as is manufacturing. Bullard *et al.* (1976) indicate that the energy intensity of automotive repairs is almost as much as that of manufacturing farm machinery, so that repair energy for some machinery is of the same order of magnitude as manufacturing energy.

Other energy consuming inputs for agricultural machinery might include taxes, shelter (if used) and insurance. These, however, are likely to be minor in comparison to manufacturing, repair and maintenance, and fuel and lubrication energy requirements.

As an example of the energy requirements for machinery, a tractor in the 100 PTO hp class, the John Deere 4240, costing $21,090, might require 0.915×10^6 MJ to deliver to the farmer. It might be operated for a total of 10,000 hours over a 14 year expected life, or each hour's use might be assigned 91.5 MJ of energy sequestered in the tractor. Estimating fuel consumption at an average 50% load, or $17.3\ L \cdot h^{-1}$, yields a fuel energy requirement of $818.3\ MJ \cdot h^{-1}$. Engine oil and filter energy

[1] Data for automobiles indicates repair costs total 6% of automobile initial costs (price). The value of 6% of initial cost as an estimate of repair costs for farm machinery was based upon automobile data.

requirements estimated as 4% of fuel is 32.7 MJ $\cdot$ h^{-1}. Repairs totaling 129% of manufacturing energy would amount to 118.0 MJ $\cdot$ h^{-1}. The total is 1060.5 MJ$\cdot$$h^{-1}$, of which 77% is for fuel.

IRRIGATION

Irrigation is provided for many different crops worldwide and with many different kinds of pumping and distribution systems. Energy is required for all irrigation systems, except that those relying totally on gravity flow may require less energy with most of the necessary energy required for equipment and facilities.

Energy is required to lift water. The quantity required ranges above the theoretical minimum of 9.81 Joules to raise a kilogram of water one meter ($9.81 \cdot J \cdot kg^{-1} \cdot m^{-1}$), or 9810 Joules to raise a cubic meter of water one meter. Considerable energy losses occur due to pipe friction, pump inefficiency, motor inefficiency, pressure drops at sprinklers or other application devices, etc. Further, water and energy losses occur because not all the water is available for the plant to use, i.e., the irrigation efficiency is less than 100%, and because the water available for plant use may not all be used.

FEA (1976) estimated United States irrigation direct energy consumption for 1974 at 0.275×10^{12} MJ. This figure included 0.020×10^{12} MJ of electric power which, when included as primary energy consumption, raises the total to an estimated 0.335×10^{12} MJ. Sloggett (1977) gave details of the FEA analysis. Faidley (1977) estimated North American direct energy for irrigation at 0.031×10^{12} MJ, or only about $^1/_{10}$ that of FEA. He estimated worldwide direct energy for irrigation at 0.147×10^{12} MJ. Faidley's assumption of 160 – 200 kg $\cdot$ ha^{-1} fuel (6900 – 8700 MJ $\cdot$ ha^{-1}) seems low, particularly for industrialized agriculture.

About 10% of United States crop acreage is irrigated. However, direct energy requirements for irrigation of 0.335×10^{12} MJ represent 30% of the total of all direct energy requirements for crop production of 1.133×10^{12} MJ. Irrigation, when used, makes agriculture much more energy intensive. Numerous examples have indicated that irrigation may raise direct energy requirements to as much as 10 times that required for agricultural machinery alone. On the other hand, irrigation enables much land to be farmed that would not be farmed otherwise.

Batty *et al.* (1975) described and modeled the energy requirements for nine types of irrigation systems:

(1) Surface irrigation system using an open-ditch network.
(2) Surface irrigation system using a gated-pipe distribution network and an Irrigation Runoff Recovery System.

(3) Solid-set aluminum lateral-pipe sprinkler irrigation system.
(4) Permanent buried plastic lateral-pipe sprinkle irrigation system.
(5) Hand-moved portable aluminum lateral-pipe sprinkle irrigation system.
(6) Side-roll aluminum lateral pipe with steel wheels sprinkle irrigation system.
(7) Center-pivot sprinkle irrigation system.
(8) Big-gun traveler sprinkle irrigation system.
(9) Trickle irrigation system.

These types typify those in industrialized agriculture. Figure 6.3 shows energy, direct and indirect, ranging from 0.000157 to 0.0136 MJ · kg^{-1}. Indirect (installation) energy when amortized over the expected life of the system and compared to direct (pumping) energy with zero lift ranged from 18% of direct energy for the traveler sprinkle to 375% for the surface with runoff recovery system.

Total diesel fuel requirements were tabulated by Johnson and Henderson (1977) for Nebraska conditions. They calculated fuel requirements ranging from 11,666 to 66,408 MJ · ha^{-1} (0.00459 to 0.00871 MJ · kg^{-1} water) for center pivot systems and 3659 to 41,953 MJ · ha^{-1} (0.00144 to 0.00551 MJ · kg^{-1} water) for gravity distribution systems.

From Sloggett's (1977) data the authors of this book calculated the United States average direct energy requirements for irrigation as being 23,629 MJ · ha^{-1} and 0.00395 MJ · kg^{-1} water applied. The installation (indirect) energy requirements, determined by Batty *et al.* (1975), ranging from approximately 1100 to 6400 MJ · ha^{-1} for nine systems, represent, therefore, only a minor portion (5% to 27%) of direct (pumping) energy costs.

Wensink *et al.* (1976) modeled seven types of irrigation systems and determined total energy requirements ranging from approximately 0.00039 to 0.0130 MJ · kg^{-1}, based upon variations in several design parameters.

LABOR

An important consideration in agricultural energy analyses is the energy sequestered in agricultural labor. Whatever method used for assigning a value, if any, is influential on results, conclusions and subsequent recommendations or actions. Whatever value is assigned will greatly affect, for example, recommendations concerning substitutions between labor and other inputs, such as in decisions relating to mechanization or automation.

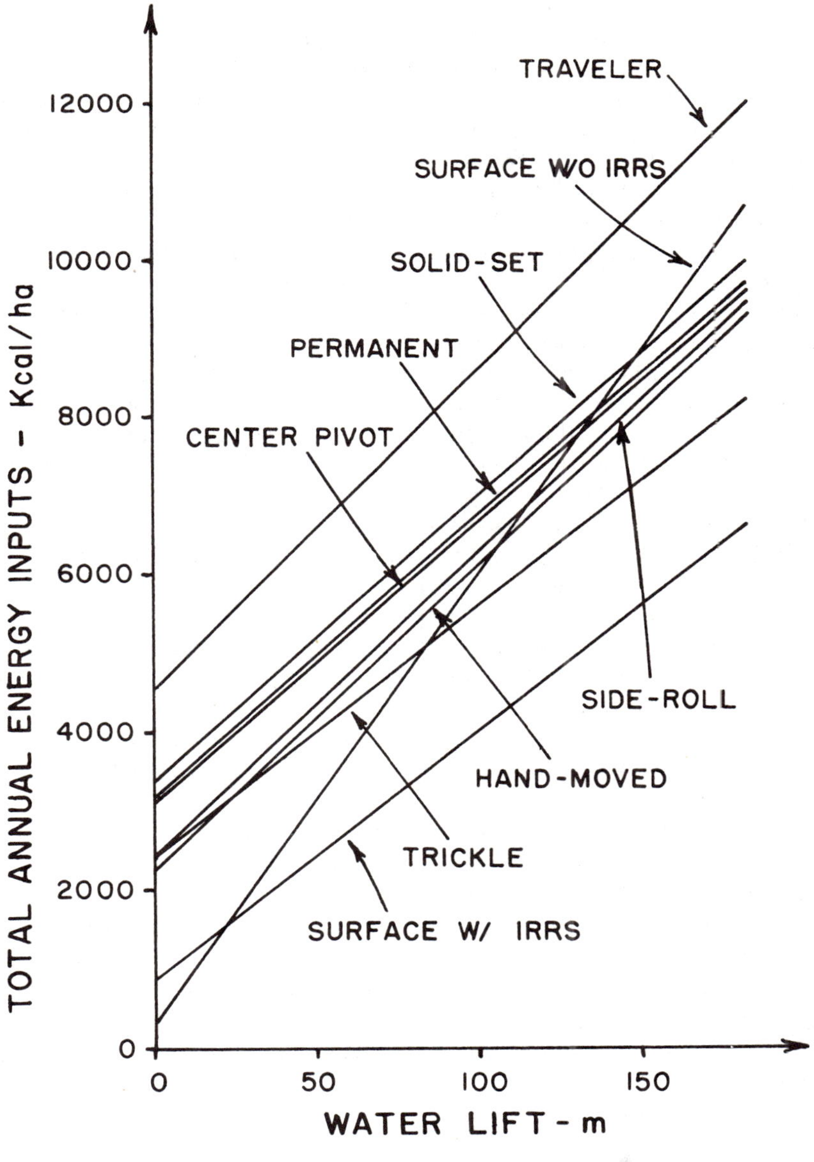

From Batty et al. (1975)

FIG. 6.3. TOTAL ANNUAL ENERGY INPUTS PER HECTARE IRRIGATED FOR NINE IRRIGATION SYSTEMS AS A FUNCTION OF WATER LIFT, BASED ON 0.915 METERS NET IRRIGATION REQUIREMENT

We will first review various methods used to assign energy values to labor, including results and rationale. We will then present a net energy analysis of the energy sequestered in agricultural labor.

An initial consideration is whether it is logical to attempt a conversion of labor to energy units. Although an energy theory of value has been called for by some (Gilliland 1975; Hannon 1973), this does not make such an attempt necessary. In fact, such a measure has been rejected by other energy analysts (IFIAS 1974; Langham and McPherson 1975; Chapman 1976). Rather, energy measures of labor have been widely used because labor may have an intrinsic energy value as muscular work and because of the energy content of food consumed by labor. More importantly, labor is substitutable to a degree (Hannon 1975; Pimentel *et al.* 1975) for other energy-consuming inputs to production systems. Despite these reasons, and perhaps because of the uncertainties involved in the presently used methods, there has been reluctance by many (Blaxter 1975; Chapman 1974,1976; Pierotti *et al.* 1977; and Webb and Pearce 1975) to apply an energy measure to labor.

Energy analysis quantifies the energy sequestered in making a good or providing a service. We extend this approach to the provision of labor as a vital input to productive activities.

There are two different methods of analyzing energy flows. One is to examine the true or thermodynamic energy flows. All forms of energy are included, conversion inefficiencies produce wastes, and ultimately all low entropy energy is utilized to yield high entropy energy. With each succeeding conversion the amount of energy remaining from a single source is less. For convenience here we shall call this method "thermodynamic."

The second, which we shall label the "sequestered" method, is that used by energy analysts who are concerned with the consumption of our stock of fossil fuels or primary energy resources. Convention allows the inclusion of the fossil fuel energy equivalent of nuclear, hydroelectric and geothermal sources, but excludes natural forms such as solar and muscular energy. The summation of all such energy required to produce a good or service is considered to be sequestered in the product or service.

Several methods which evaluate agricultural labor energy will be presented as follows in order of increasing energy levels obtained, and will be identified with the thermodynamic and sequestered methods just described.

Muscular Energy Expended by Labor.—Makhijani (1975) used a range of 0.49–0.58 $MJ \cdot d^{-1}$ for net energy output (with 8.44 $MJ \cdot d^{-1}$ food intake) of agricultural labor in his examination of Third World agriculture. This is the true or thermodynamic energy value of muscular work expended. The ultimate quantity and source of the energy, whether fossil fuel, solar or otherwise, used to supply this energy is not of immediate concern. This method will likely be of minimum usefulness in cases where much time is spent in management or decision-making activities rather than in manual labor.

Portion of the Energy Content of Consumed Food.—The apportionment is generally based on time, perhaps weighted by metabolic rate, among daily activities. Gifford and Millington (1975) used 7.1 MJ · d^{-1}, Revelle (1976) used 6.4 and 8.0 MJ · d^{-1} for women and men, respectively, on an average daily basis over a year, and Black (1971) used 5 MJ · d^{-1}. This, also, is a thermodynamic measure of the energy of labor but with the energy input counted as food caloric energy one conversion (metabolic) previous to the muscular energy application. Passmore and Durnin (1955) reviewed the food energy requirements for various activities.

Total Energy Content of Consumed Food.—This is the method most widely used (Batty *et al.* 1975; Dekkers *et al.* 1974; Hudson 1975; Keener and Roller 1975; Newcombe 1975; Pimentel *et al.* 1973; Stanhill 1974) to designate the energy content of labor. Values used the range from 8.0 to 18.2 MJ · d^{-1}. This also is a thermodynamic measure, but in this case no reduction is made for other uses to which a portion of the food energy may be put, such as leisure time activities.

None of these first three methods is a measure of the primary energy resources consumed in providing the labor. Although having value otherwise, none of the three should be used by the energy analyst whose interest is in quantifying those energy resources consumed. The following methods are based upon the depletion of primary energy resources.

Farm Family Support Energy.—Williams *et al.* (1975) presented values of household energy uses from which can be obtained the value 607.8 MJ · d^{-1}. ERS (1974) estimated farm family fuel use from which we obtain 414.6 MJ · d^{-1}. Both values are for United States agriculture and include only direct energy consumption. Neither source presented these values as being the energy value for labor, but they give a preliminary idea of the magnitude of the energy which might be sequestered in labor. Of course, some of this energy is consumed providing for leisure time pursuits also.

Marginal Substitution Ratio.—De Wit (1975) developed a theory of labor-energy substitutions in agricultural production. Hyperbolic isoyield curves were drawn on "added labor" and "added energy" axes which sum respectively the labor and energy sequestered in the product.

A specific point on an iso-yield curve represented a specific production situation and indicated the labor and energy sequestered in the product when produced under that production situation. The slope of the iso-yield curve at a point defined the marginal substitution ratio. De Wit found, for the one example of modern Dutch agriculture he presented, a marginal substitution ratio of 190 GJ $\cdot$ man^{-1} $\cdot$ yr^{-1}. If one assumes 230 working days per man year, the marginal substitution ratio is 826 MJ $\cdot$ man^{-1} $\cdot$ d^{-1}. This value represents, for example, the rate at which energy is substituted for labor when a more energy intensive production system is substituted for a less energy intensive one. De Wit and Van Heemst (1976) subsequently stated that the present marginal substitution ratio is 125 GJ $\cdot$ man^{-1} $\cdot$ yr^{-1}; this is 543 MJ $\cdot$ man^{-1} $\cdot$ d^{-1}.

Lifestyle Support Energy.—This is the total energy sequestered in the services consumed by the farmworker and perhaps also by his/her family. The energy involved is that energy sequestered in the goods and services purchased by the wages earned from working. This approach is a sequestered energy method. Slesser (1973) termed this energy "life support energy" and gave data from which one can infer its value to be at least 229 MJ $\cdot$ d^{-1}. Avlani and Chancellor (1977) used the term "lifestyle support energy," but made no evaluations. Jones (1975) and Hawthorne (1975) both argued for this approach, and values from 700 to 1400 MJ $\cdot$ d^{-1} can be inferred from Jones' value of 40 MJ $\cdot$ ha^{-1}. DeBellevue (1976), in his vegetable crop model, used 2102 MJ $\cdot$ d^{-1} (8 hr $\cdot$ d^{-1}) for all cultural labor and 2473 MJ $\cdot$ d^{-1} for picking labor. Fluck (1976) presented data and suggested a procedure for calculating this energy based upon the GNP; for the average United States laborer working in 1970 it was 3600 MJ $\cdot$ d^{-1}.

To evaluate the lifestyle support energy value of labor as a per capita portion of the energy sequestered in providing the goods and services in the GNP, however, is to count twice the energy sequestered in all goods and services produced. That is, all energy is counted once as direct and indirect energy inputs to producing the goods and services represented by the GNP. The GNP is consumed by or in behalf of the population, and labor supplied from that population evaluated in terms of the energy sequestered in the GNP counts the energy a second time as a labor input. This energy recycling or double accounting is fallacious. Bullard (1975) and Avlani and Chancellor (1977) recognized this error.

The first IFIAS energy analysis workshop (1974), which supported a life support or sequestered energy value rather than a food energy or thermodynamic value, in keeping with the established procedures for energy analysis, apparently did not recognize the double accounting involved. Neither did Fluck (1976) nor, apparently, DeBellevue (1976).

The workshop left unresolved the question of partition of life support energy between labor and nonlabor or, as nonlabor was termed, "household." The workshop did, however, recommend that life support energy need not be included in energy analyses of industrialized economies, presumably because it was thought to be small in comparison with other energy inputs.

Net Energy Analysis of the Energy Value of Labor.—The net energy analysis of the energy value of labor was devised to estimate the energy sequestered in labor and to partition the total lifestyle support energy between labor and nonlabor or "household" uses. It is based on net energy analysis described by Odum and Odum (1976) and Bullard (1975) in which energy-producing industries and energy conservation practices are analyzed for effectiveness, recognizing that a portion of the energy obtained or saved is required as feedback energy to provide the activities and facilities for energy extraction or conservation. Analogously, for labor, a portion of the goods and services consumed and energy sequestered in them is required as feedback to enable the laborer to pursue his/her wage earning activities. The remainder is support for his/her family, leisure time activities, etc. Readily identifiable goods and services directly supporting an individual's wage-earning activities include a portion of his/her total consumption of food, clothing, transportation, housing, etc. The remaining energy sequestered in other consumed goods and services is that net output which supports family, luxuries, leisure time activities, retirement, etc. Figure 6.4 illustrates the concept of the net energy analysis of the energy sequestered in labor.

Justification for the net energy analysis of the energy value of labor is dependent to a degree upon one's answer to a philosophical question. That is, is work viewed by the worker as a means for providing a livelihood for him and herself and his/her family, or is it viewed as a desirable activity in itself? A spectrum of responses to this question obviously exists, but this model assumes that productive employment occurs in order that other activities might be enjoyed. Further, no feedback is associated with the production of goods and services not included with the GNP. Examples of the latter would include household repairs and maintenance, home meal preparation and home child care for which there exists no employee-employer relationship. The position reflected here is that employment is sought to provide a livelihood which includes the provision of goods and services for leisure time activities.

Details of the methodology of the net energy analysis of the energy value of labor are given by Fluck (1979). The procedure involves a disaggregation of the GNP into its components and estimation of the energy content and portion of each consumed as feedback to provide

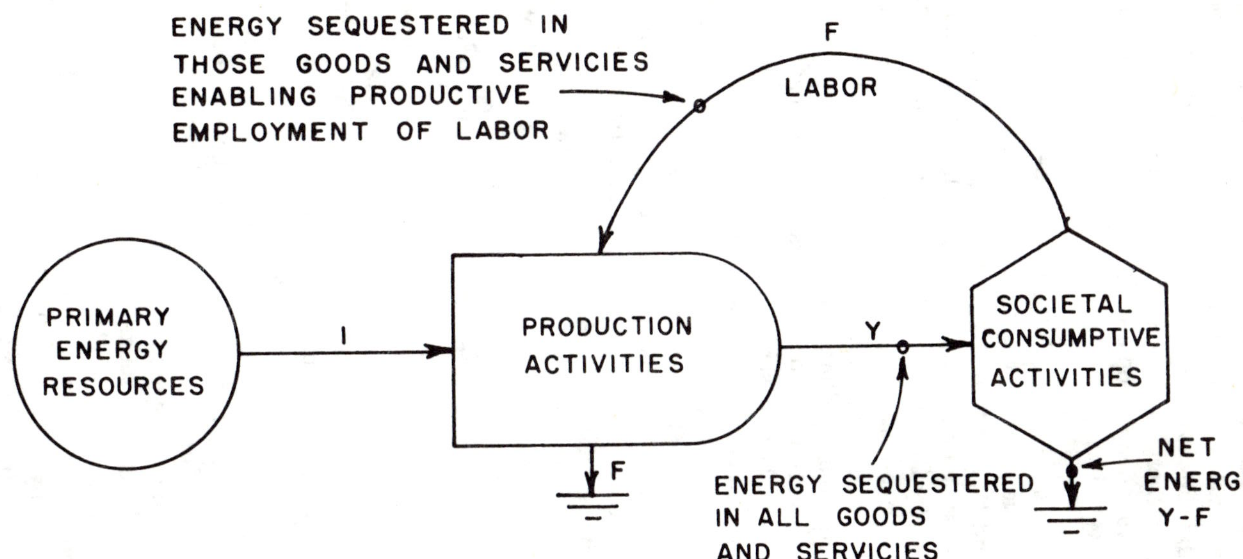

FIG. 6.4. NET ENERGY ANALYSIS OF THE ENERGY SEQUESTERED IN LABOR

Energy supplied from energy sources is I, the energy sequestered in GNP is Y, the energy sequestered in labor is F, and the net energy sequestered in foods and services not required for labor is $Y - F$. Since this is a sequestered energy model, $Y = I$.

labor. Included within GNP are the four categories: personal consumption expenditures, governmental purchases of goods and services, gross private domestic investment and net exports. Personal consumption expenditures are by definition consumed by people. Gross private domestic investments include personal housing and investments not made by individuals but by private industry which benefits individuals through the ultimate provision of personal consumption expenditures. Government purchases are made for the benefit of the individuals governed. Net exports represent yearly fluctuations from the anticipated long term balance of exports and imports.

Figure 6.5 summarizes results of the net energy analysis model for the 1974 United States economy. The feedback energy of $16{,}968 \times 10^9$ MJ providing labor was 22.2% of the total. If the annual labor feedback energy of $16{,}968 \times 10^9$ MJ is divided by the number of persons in the labor force (88.1×10^6) and by an assumed number of days worked (230), the energy sequestered in the average person employed in the United States in 1974 was 837 $MJ \cdot d^{-1}$.

This average value for all labor of 837 $MJ \cdot d^{-1}$, is somewhat high for agricultural labor. Herendeen and Tanaka (1975) found the energy intensity of rural farm household expenditures was about 23% above that for urban nonfarm households. However, the earnings of farmers, farm managers and farm laborers are about 42% less (U.S. Bur. of Census 1976) than those of nonagricultural employees. The net effect of these two phenomena is a reduction of approximately 29%. Therefore, the energy sequestered in all United States agricultural labor in 1974 is estimated to be 594 $MJ \cdot d^{-1}$.

The net energy analysis value of 594 $MJ \cdot d^{-1}$ for the energy sequestered in agricultural labor is approximately 45 times the value heretofore most widely used, 13 $MJ \cdot d^{-1}$, based on food caloric energy consumption. The net energy analysis model gives an estimate of the primary energy consumed in order to provide agricultural labor, whereas other methods do not. It is the preferable method, therefore, to assign energy values to labor in energy analyses whose purposes include evaluation of the consumption of primary energy.

Use of the net energy analysis model results in much greater energetic importance being placed upon labor. Instead of being insignificantly small, labor energy becomes an item of some consequence. In an energy analysis of staked tomato production (Fluck *et al.* 1978), labor accounted for 43% of the total production and harvesting costs and, using the net energy analysis model to evaluate labor energy, 7% of the total energy requirements. If labor energy had instead been evaluated at 13 MJ $\cdot d^{-1}$, it would have accounted for only about 0.3% of the total energy requirements.

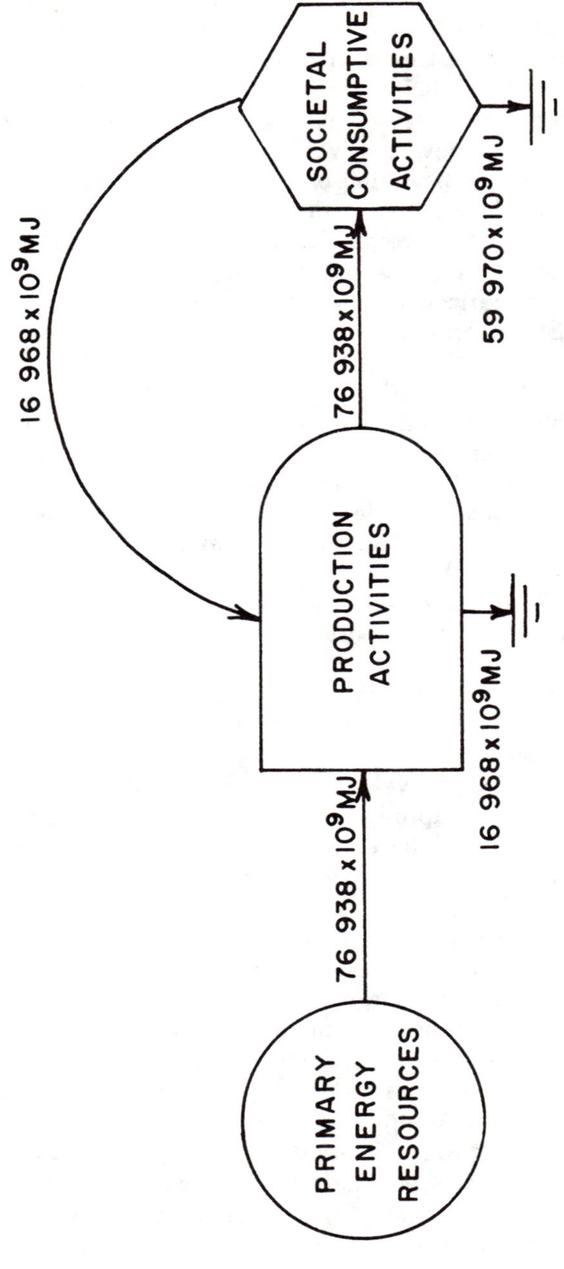

FIG. 6.5. NET ENERGY EVALUATION OF THE ENERGY SEQUESTERED IN LABOR FOR THE 1974 UNITED STATES ECONOMY

LAND

Energy flows are associated with land and the ownership of land. To collect and utilize solar energy requires area on the surface of the earth. Energy flows are associated with the costs of land use or ownership. Capital improvements to land require energy. Land and other energy-requiring inputs are generally partially substitutable one for another (Chancellor and Goss 1976). Therefore, for various purposes, it may be desirable or necessary to quantify land in terms of energy.

The average value of United States farmland (February 1977) is \$1128 $\cdot$ ha^{-1}. Ownership or rent of farmland costs the user, either in rent, mortgage payments or lost opportunity for other investments. A minimal cost for the average United States farmland, paid in any one of the above three forms, would be about \$90 $\cdot$ ha^{-1} $\cdot$ yr^{-1}. The average rent for United States agricultural land is now about \$260 $\cdot$ ha^{-1} $\cdot$ yr^{-1}. For ownership, an additional item is property taxes; property taxes on farm real estate averaged \$7.22 $\cdot$ ha^{-1} $\cdot$ yr^{-1} in 1975 (USDA 1977). Much agricultural land is changed from its natural state with such improvements as clearing, leveling, terracing, and irrigation and drainage. These capital improvements can cost as much as several thousand dollars per hectare, but normally the costs of such improvements are included in the value of the land. An estimated total annual user cost of \$100 $\cdot$ ha^{-1}, when multiplied by the average 1975 United States GNP energy intensity of 49.89 MJ $\cdot$ \$$^{-1}$, yields an energy cost of 4989 MJ $\cdot$ ha^{-1} $\cdot$ yr^{-1}. Alternatively, we have from Bullard *et al.* (1976) an energy intensity for real estate of 20.53 MJ $\cdot$ \$$^{-1}$, yielding an energy cost of 2053 MJ $\cdot$ ha^{-1} $\cdot$ yr^{-1}, likely a better estimate reflecting the lower energy intensity of transactions involving real estate. It is further likely that agricultural real estate may be less energy intensive than other real estate, justifying a somewhat lower energy cost than 2053 MJ $\cdot$ ha^{-1} $\cdot$ yr^{-1}.

The justification for an energy value for land based on the solar energy input was stated by Heichel (1976): "Traditionally, the sun's energy has been considered a 'free good'. However, the farmer must own or rent land to support the plants that capture sunlight, and the increasing price of farmland directly reflects the increasing value of photosynthetically active sunlight." The value of agricultural land depends in part on its ability to derive an income using solar energy inputs. Georgescu-Roegen (1969) termed the land a "catching net" for not only solar energy but also "chemicals" such as carbon dioxide and, presumably, water in the form of precipitation.

The total annual solar radiation on an average United States location is in the order of 76×10^6 MJ $\cdot$ ha^{-1}. If this amount is divided by 2000 as suggested by Odum and Odum (1976) to convert to fossil fuel equivalents, we have an estimated natural energy flow of solar energy on land of about 36,000 MJ · ha^{-1} · yr^{-1}. Rainfall is a component of the natural energy flow. Burnett (1978) has estimated the fossil fuel equivalent energy of rainfall at 0.00472 MJ · kg^{-1}. The average United States location which receives an annual rainfall of approximately 1 m, then receives a natural energy flow in the form of rainfall of 23,600 MJ · ha^{-1} · yr^{-1}. Rainfall results from solar energy and should not be added to solar to avoid double counting.

Regan (1977) determined the energy equivalent of three Florida soils based upon the natural energy flows required for soil formation. Expressed as fossil fuels they were 6.79×10^6 MJ · ha^{-1} and 9.02×10^6 MJ · ha^{-1} for mineral soils and 19.65×10^6 MJ · ha^{-1} for peat soil. If it is assumed the soils are depleted with, for instance, 1000 years use, the soil formation energies range from 6790 to 19,650 MJ · ha^{-1} · yr^{-1}.

Thus we have several different estimates of the energy flows associated with land, ranging from 2000 MJ · ha^{-1} · yr^{-1} to 36,000 MJ · ha^{-1} · yr^{-1}. Each may be useful depending upon individual circumstances.

CONTAINERS

Agricultural products are generally held within some container after harvest and during transport to market. Container size varies from as small as approximately 0.001 m^3 to truck bodies of 60 m^3. Farm containers tend to be large, durable and reusable. They include bags, baskets, crates, boxes, pallet bins and bulk truck and trailer bodies.

Energy intensities (MJ · $) can be obtained from the data of Bullard *et al.* (1976) for several types of containers. The energy intensities given in Table 6.4 are in megajoules per 1974 dollar.

Energy requirements for containers can be quantified in terms of energy per unit weight or volume. Reusable containers might be quantified in terms of energy per unit weight or volume contained per use. Table 6.5 presents estimates of the energy requirements for provision of several agricultural containers.

TRANSPORTATION

Estimates of the direct energy inputs for transporation in the United States at the farm level in 1974 were made by FEA (1976). Energy inputs for trucks and pickups were 0.250×10^{12} MJ and for autos were 0.085×10^{12} MJ for a total of 0.335×10^{12} MJ. Indirect energy inputs are likely

TABLE 6.4. ENERGY INTENSITIES OF CONTAINERS

Container	Energy Intensity (MJ · 1974 $$^{-1}$)
Wooden containers	34,339
Paperboard containers and boxes	72,588
Glass containers	86,530
Metal cans	91,767
Fabricated textile products (cloth and burlap bags)	47,828

Source: Bullard *et al.* (1976).

TABLE 6.5. ENERGY REQUIREMENTS FOR CONTAINERS

Container	Cost ($)	Energy per unit volume contained (MJ · m^{-3})
Mesh bag	0.035	3150
Cloth or burlap bag	0.25	180
Corrugated cardboard box	0.35	960
Wire bound crate	0.75	780
Nailed wood crate		
Single use	6.00	3890
10 uses	6.00	389
Pallet box		
Single use	25.00	1300
20 uses	25.00	65
Bulk truck body		
Single use	5000.00	5582
100 uses	5000.00	56

in the order of half or less this amount for a total annual United States agricultural transporation energy requirement in the order of 0.5×10^{12} MJ. It is obvious that a significant portion (probably at least 25%) of the agriculturally-related transportation is of people rather than other inputs or outputs. This estimate does not include transportation of inputs through the manufacturing and distribution system nor of agricultural products from the point where farm transportation ceases.

Energy requirements for freight hauling vary with mode and speed. Tucker (1975) indicated that there exists for each mode an optimum speed at which energy costs are minimized. Increased energy costs occur when speeds are increased above this optimum point.

Energy requirements for freight hauling are normally given in energy intensiveness, the energy units per unit of weight per unit of distance traveled. We will use MJ · t^{-1} · km^{-1}(t = tonne = 1000 kg).

Table 6.6 lists the energy intensiveness determined from several sources for five transportation modes. Some are for direct energy only whereas several also include indirect energy inputs.

Direct energy costs probably constitute the majority of total energy costs for all modes of transportation. Fluck and Shaw (1976) found, for owner operator tractor-trailer trucks, that 76% of the total energy requirements were for the direct energy inputs of fuel and oil. Indirect energy requirements were 7% of total for taxes and licenses, 6% for depreciation, 3% for repairs and service, 3% for tires and tubes, 2% for insurance and safety and 3% for other.

Energy intensities for various modes of transportation are listed in Table 6.7.

DRAFT ANIMALS

Horses, mules, oxen, bullocks and other draft animals are an important source of power in much of the world's agriculture. A single draft animal increases by a factor of about 10 the power available to a man. Draft animals have been largely replaced by tractors in industrialized agriculture. Faidley (1977) estimated that draft animals supply 25% of the traction power in developing countries ranging from a much higher percentage in Asia to almost nonexistent in the Tsetse fly belt of Africa.

TABLE 6.6. ENERGY INTENSIVENESS (EI) OF MATERIALS TRANSPORTATION

Mode	EI ($MJ \cdot t^{-1} \cdot km^{-1}$)	Source
Air	30.3	Hirst (1973)
	8–50	Rice (1972)
Truck	2.02	Hirst (1973)
	0.56–4.52	Leach and Slesser (1973)
	1.34–4.18*	Penner (1974)
	1.65–4.52*	Penner (1975)
	2.67	Rice (1972)
Rail	0.48	Hirst (1973)
	0.30	Leach and Slesser (1973)
	0.20–0.72*	Penner (1974)
	0.96*	Sebald (1974)
	0.18	Rice (1972)
Water	0.49	Hirst (1973)
	0.05–0.12	Leach and Slesser (1973)
	0.27–0.84*	Penner (1974)
	1.18	Sebald (1974)
	0.10–0.67	Rice (1972)
Pipeline	0.31	Hirst (1973)
	0.17	Rice (1972)

*Direct and indirect energy

TABLE 6.7. ENERGY INTENSITY FOR VARIOUS MODES OF TRANSPORTATION

Mode	Energy Intensity (MJ · 1974 $^{-1}$)
Air (freight and passenger)	177
Truck	48
Rail (freight and passenger)	68
Water	233
Pipeline	32

Source: Bullard *et al.* 1976).

In the United States the number of horses and mules peaked about 1918 at 27 million and has steadily decreased since to fewer than five million currently. Only a small percentage of the current number could be classified as draft animals.

Spedding (1975) listed draft (pulling or traction force) of several animals and also load carrying capacities. Draft ranged from 45–135 kg for reindeer to 1000 kg for Chinese humpless cattle. Load carrying capacities ranged from 40 kg for reindeer to 250 kg for the camel.

The work performed when a draft animal pulls a load is the product of draft or pull and distance. The work rate, or power, is the work performed per unit time or the product of draft and speed.

Data presented by Brody (1974) are comprehensive sources of bioenergetic information on draft animals. He defined several efficiences:

$$\text{Gross eff.} = \frac{\text{work accomplished}}{\text{energy expended}}$$

$$\text{Net eff.} = \frac{\text{work accomplished}}{\text{energy expended above that expended at rest}}$$

$$\text{Absolute eff.} = \frac{\text{work accomplished}}{\text{energy expended above that of walking without load}}$$

Maximum efficiencies observed for horses (Brody 1974) have been 25% for gross efficiency, 28% for net efficiency and 35% for absolute efficiency. Typical efficiencies are considerably less. Efficiency varies with speed (from low to a maximum at approximately 5–6 km · h^{-1}). All-day efficiency is considerably less than the instantaneous efficiences above and varies with amount of time worked per day.

Brody (1974) made a tentative recommendation of a work rate for horses to maximize efficiency. He recommended a draft of 10% of body

TABLE 6.8. ALL-DAY EFFICIENCIES FOR HORSES

Hours worked per day	Work accomplished (MJ)	Energy expended (MJ)	All-day efficiency (%)
12	32.2	195.1	16.5
10	26.9	172.9	15.5
8	21.5	150.3	14.3
6	16.1	128.1	12.6
4	10.8	105.5	10.2
2	5.4	83.3	6.4
0	0.0	61.1	0.0

Source: Spedding (1975).

weight at 4 $km \cdot h^{-1}$. At this draft and speed a horse of 680–860 kg can work 10 $h \cdot d^{-1}$ at 2.685 $MJ \cdot h^{-1}$ (1 hp).

Feed for draft animals may come from sources that do not provide human food, as when the feed is forage grown on land unsuitable for cultivation. Where land is plentiful, draft animals may graze cultivable land and not compete with humans. However, where human populations require most or all of the cultivable land and where noncultivable land is insufficient, draft animals are few. Draft animals require feed even when not working and require large land areas for feed production. Gavett (1973), to emphasize the infeasibility of a return to draft animals by industrialized agriculture, stated that 61 million draft animals would be required to power today's United States agriculture and that these draft animals would consume the feed produced from over 73×10^6 ha of cropland, or almost half the United States cropland.

In order to determine the energy requirement for draft animals, detailed examination must be made of the systems supplying feed for the animals. Some require primary energy; others in lesser developed countries may require little or no primary energy but instead rely on natural energy flows. It is recommended that an accounting of support energy not focus on the caloric energy of the feed consumed but rather on the energy requirements for producing that feed and all other animal inputs. Further, it is recommended that all inputs be included over the draft animal's life rather than including only those during the working season or during working hours.

Comparisons of the costs of tractors and bullocks in Indian agriculture were made by Bhatia (1977) and Rao and Singh (1977). In both studies bullocks were found to be less expensive than tractors.

Draft animals may have significant outputs in addition to muscular work. Odend'hal (1972) studied Indian cattle. Their feed input consisted mainly of rice straw with lesser quantities of mustard oil cake, wheat bran, rice hulls, chopped banana tree trunks, and grazing on canal banks, roadsides, paths and railroad tracks. Total energy input per head was 60.33 MJ · d^{-1}. Useful outputs per head consisted of work of 1.71 MJ · d^{-1} (average work rate for bullocks was taken at 1.81 MJ · h^{-1}), dung at 11.96 MJ · d^{-1}, milk at 0.56 MJ · d^{-1} and calves at 0.03 MJ · d^{-1}. Total outputs were 14.26 MJ · d^{-1} and overall efficiency assuming 67% of the dung was used as fuel was 16.9%.

ENVIRONMENTAL CONTROL

Significant quantities of energy are consumed in agriculture to modify the environment for growing livestock and plants, and storing agricultural products. Energy is used for heating buildings for young livestock in colder climates and in colder seasons, for cooling livestock by ventilation, for providing refrigeration for preservation of products, for heating greenhouses, and for frost protection. FEA (1976) estimated 0.045×10^{12} MJ was used in United States agriculture in 1974 for frost protection, and 0.039×10^{12} MJ for livestock space heating, ventilation and brooding of young livestock. CAST (1975) listed the 1969 United States energy consumption for greenhouse heating at 0.041×10^{12} MJ.

The amounts of energy required for a specific environmental control application vary with the specifics of structural design, weather and management.

ON-FARM PROCESSING AND HANDLING

Included in the above category are crop drying (0.111×10^{12} MJ), handling of feeds, livestock, water, etc. (0.099×10^{12} MJ), and heating and cooling processes (0.015×10^{12}), each as determined by FEA (1976) for the United States for 1974.

Crop drying is basically evaporation of water with theoretical minimum energy requirements at approximately 2.50–2.67 MJ · kg^{-1} water, depending on the temperature at which water is evaporated. Actual energy requirements for different types of grain driers used for drying corn were found to range from 3.45–4.14 MJ · kg^{-1} of water removed (Meiering *et al.* 1977). Maddex and Bakker-Arkema (1978) stated that energy requirements for evaporating water from grain range from 3.02 to 6.98 MJ · kg^{-1} of water. Morey *et al.* (1976) found for a continuous crossflow drier energy, requirements beginning in the range of 4.3–5.6 MJ · kg^{-1} of water removed but increasing with less than optimum management (higher air flow rates, lower air temperatures) to 8–14 MJ · kg^{-1} of water removed.

Energy used to operate conveyors varies considerably with type, e.g., the energy required to operate a pneumatic conveyor might be several times that required to operate a belt conveyor or bucket elevator.

OTHER (PLASTIC MULCHES, ETC.)

Farmers purchase numerous other items as inputs to agricultural production. Examples of additional industry-supplied inputs are binder twine, baling wire, plastic mulch, fencing materials, hardware and small tools, paint, dairy supplies, nursery and greenhouse supplies, etc. Little information is currently available on the energy requirements for many of these inputs. Estimates of energy requirements can be obtained using the ratio of national energy consumption to GNP or input-output analysis coefficients.

As an example, the energy sequestered in binder twine is estimated. From Bullard *et al.* (1976) we obtain an estimate of the energy intensity of cordage and twine of 53.22 MJ $\cdot$ $\$^{-1}$. The U.S. Department of Agriculture (1977) gives the 1974 price paid by farmers as \$1.49 $\cdot$ kg^{-1}. The product is 79.2 MJ $\cdot$ kg^{-1}.

As another example, the energy sequestered in black plastic (polyethylene) mulch was determined by Fluck *et al.* (1978) by process analysis to be 158 MJ $\cdot$ kg^{-1}.

Farmers also purchase considerable quantities of agricultural products from other farmers and industry. Feed, livestock and seed constitute about 30% of all production expenses.

BIBLIOGRAPHY

AMER. SOC. AGRIC. ENG. (ASAE). 1978. Agricultural machinery management, ASAE *EP391*. *In* 1978–79 Agricultural Engineers Yearbook. Amer. Soc. Agric. Eng., St. Joseph, Michigan.

ANON. 1978. National Farm Tractor and Implement Blue Book. Vol. 39. Natl. Mrkt. Rept., Chicago.

ANON. 1979. Nebraska Tests Section. Implement and Tractor *93* (3) A-141—A-177.

AVLANI, P.K. and CHANCELLOR, W.J. 1977. Energy requirements for wheat production and use in California. Trans. ASAE *20,* 429–437.

BATTY, J., HAMAD, S.N. and KELLER, J. 1975. Energy inputs to irrigation. J. Irr. and Drainage Div., ASCE. *101* (IR4) 293–307.

BERRY, R.S., FELLS, M.F. and MAKINO, H. 1973. A thermodynamic valuation of resource use: Making automobiles and other processes. *In* Energy:

Demand, Conservation and Institutional Problems. Mass. Inst. Technol. Press, Cambridge, Mass.

BHATIA, R. 1977. Energy and rural development in India: Some issues. *In* Energy and Agriculture. W. Lockeretz (Editor). Academic Press, New York.

BISWAS, A.K. and BISWAS, M.R. 1975. Energy and food production. Agro-Ecosystems *2* (3) 195–210.

BLACK, J.N. 1971. Energy relations in crop production—A preliminary survey. Ann. Appl. Biol. *67,* 272-278.

BLAXTER, K.S. 1975. The energetics of British agriculture. J. Sci. Food Agric. *26,* 1055–1064.

BLOUIN, G.M. and DAVIS, C.H. 1976. Energy requirements for the production and distribution of chemical fertilizers in the United States. *In* Energy and Agriculture. Southern Regional Educational Board, Atlanta.

BRODY, S. 1974. Bioenergetics and Growth. Hafner Press, a division of Macmillan Publishing Co., New York.

BULLARD, C.W., III. 1975. Energy costs, benefits and net energy. CAC Doc. No. *174,* Cntr. for Adv. Computation, Univ. of Ill., Urbana.

BULLARD, C.W., PENNER, P.S. and PILATI, D.A. 1976. Energy analysis handbook. CAC Doc. No. *214* Cntr. for Adv. Computation. Univ. of Ill., Urbana.

BURNETT, M.S. 1978. Energy analysis of intermediate technology agricultural systems. M.S. Thesis, Univ. of Fla., Gainesville.

CERVINKA, V. *et al.* 1975. Methods used in determining energy flows in California agriculture. Trans. ASAE *18* (2) 246–251.

CHANCELLOR, W.J. and GOSS, J.R. 1976. Balancing energy and food production. 1972–2000. Science *192,* 213–218.

CHAPMAN, P.F. 1974. Energy costs: A review of methods. Energy Policy *2* (2) 91–103.

CHAPMAN, P.F. 1976. Principles of energy analysis. *In* Aspects of Energy Conversion. J.M. Blair *et al.* (Editors). Pergamon Press, New York.

COBLE, C.G. and LEPORI, W.A. 1974. Energy consumption, conservation and projected needs for Texas agriculture. Report S/D-12, Agric. Eng. Dep., Texas A & M Univ., College Station.

COMMONER, B., GERTLER, M., KLEPPER, R. and LOCKERETZ, W. 1974. The effect of recent price increases on field crop production costs. CBNS-AE-1, Cntr. for Biol. of Nat. Sys., Wash. Univ., St. Louis.

COUNC. AGRIC. SCI. TECHNOL. (CAST). 1975. Potential for Energy Conservation in Agricultural Production. Rept. No. *40,* Counc. for Agric. Sci. and Technol. Dep. of Agronomy, Iowa State Univ., Ames.

DAVIS, C.H. and BLOUIN, G.M. 1976. Energy consumption in the U.S. chemical fertilizer system from the ground to the ground. Div. of Chem. Dev., Tenn. Valley Authority, Muscle Shoals, Alabama.

DEBELLEVUE, E.B. 1976. Energy basis for an agricultural region: Hendry County, Florida. M.S. Thesis, Univ. of Fla., Gainesville.

DEKKERS, W.A., LANGE, J.M. and DE WIT, C.T. 1974. Energy production and use in Dutch agriculture. Neth. J. Agric. Sci. *22* (2) 107–118.

DE WIT, C.T. 1975. Substitution of labour and energy in agriculture and options for growth. Neth. J. Agric. Sci. *23* (2) 145–162.

DE WIT, C.T. and VAN HEEMST, H.D.J. 1976. Aspects of agricultural resources. *In* Chemical Engineering and a Changing World. W.T. Koetsier (Editor). Elsevier Scientific Publishing Co., Amsterdam.

DOERING, O.C., III, CONSIDINE, T.J. and HARLING, C.E. 1977. Accounting for tillage equipment and other machinery in agricultural energy analysis. NSF/RA-770128. Agric. Exp. Sta., Purdue Univ., West Lafayette, Indiana.

EDMINSTER, T.W. 1975. Weeds, energy and you. Address at 15th Meet. of Weed Sci. Soc. of Amer., Washington, D.C.

ECONOMIC RES. SERV. (ERS). 1974. The U.S. food and fiber sector: Energy use and outlook. Economic Res. Serv.—U.S. Dep. Agric., Washington, D.C.

FAIDLEY, L.W. 1977. Energy requirements and efficiency for crop production. ASAE Ppr. *77–5528* presented at Winter Meet. Amer. Soc. of Agric. Eng., St. Joseph, Michigan.

FED. ENERGY ADMIN. (FEA). 1976. Energy and U.S. agriculture: 1974 data base. FEA/D-76/459. Fed. Energy Admin.—U.S. Dep. Agric., Washington, D.C.

FLUCK, R.C. 1976. To evaluate labor energy in food production. Agric. Eng. *57* (1) 31–32.

FLUCK, R.C. 1979. Net energy analysis of the energy sequestered in agricultural labor. (Paper submitted for publication to Energy Policy)

FLUCK, R.C. and SHAW, L.N. 1976. Energy and economic comparisons of native and imported peat. Proc. Fla. State Hort. Soc. *89,* 309–311

FLUCK, R.C., SHAW, L.N. and EVERETT, P.H. 1978. Energy analysis of the use of full-bed plastic mulch on vegetables. Proc. Fla. State Hort. Soc. *90,* 382–385.

GAVETT, E.E. 1973. Agriculture and energy use. Ppr. presented at Ann. Meet. Amer. Agric. Econ. Assoc., August 8–11. Edmonton, Alberta, Canada.

GEORGESCU-ROEGEN, N. 1969. Process in farming versus process in manufacturing: A problem of balanced development. *In* Economic Problems of Agriculture in Industrial Societies. U. Papi and C. Nunn (Editors). Macmillan, New York.

GIFFORD, R.M. and MILLINGTON, R.J. 1975. Energetics of agriculture and food production. Bull. No. *288,* Commonwealth Sci. Ind. Res. Organ., Melbourne.

GILLEY, J.R. and WATTS, D.A. 1976. Energy reduction through improved irrigation practices. *In* Agriculture and Energy. W. Lockeretz (Editor). Academic Press, New York.

GILLILAND, M. 1975. Energy analysis and public policy. Science *189,* 1051–1056.

GREEN, M.B. 1976. Energy in agriculture. Chem. and Ind. *15,* 641–646.

GREEN, M.B. 1978. Eating Oil. Westview Press, Boulder, Colorado.

GREEN, M.B. and MCCULLOCH, A. 1976. Energy considerations in the use of herbicides. J. Sci. Food Agric. *27* (2) 95–100.

HANNON, B.M. 1973. An energy standard of value. Ann. Amer. Acad. Pol. Soc. Sci. *410,* 139–153.

HANNON, B. 1975. Energy conservation and the consumer. Science *189,* 95–102.

HAWTHORNE, J. 1975. Energy usage in food processing and distribution. Span *18* (1) 15–16.

HEICHEL, G.H. 1976. Agricultural production and energy resources. Amer. Sci. *64,* 64–72.

HERENDEEN, R.A. and TANAKA, J. 1975. Energy cost of living. CAC Doc. No. *171.* Cntr. for Adv. Computation, Univ. of Ill., Urbana.

HIRST, E. 1973. Energy intensiveness of passenger and freight transport modes: 1950–1970. ORNL-NSF-EP-44. Oak Ridge Natl. Lab., Oak Ridge, Tenn.

HOEFT, R.G. and SIEMENS, J.C. 1975. Energy consumption and return from adding nitrogen to corn. Ill. Res. *17* (1) 10–11.

HUDSON, J.C. 1975. Sugarcane: Its energy relationships with fossil fuels. Span *18* (1) 12–14.

INT. FED. INST. ADV. STUD. (IFIAS). 1974. Energy analysis. Workshop Report No. *6.* Int. Fed. Inst. for Adv. Stud., Stockholm.

JOHNSON, B.B. and HENDERSON, P.A. 1977. Energy price levels and the economics of irrigation. Dep. of Agric. Econ. Rept. No. *79A.* Univ. of Neb., Lincoln.

JONES, D. 1975. The energy relations of pesticides. Span *18* (1) 20–22.

KEENER, H.M. and ROLLER, W.L. 1975. Energy production by field crops. ASAE Paper No. *75-3012.* Presented at Ann. Meet. Amer. Soc. Agric. Eng., St. Joseph, Michigan.

LANGHAM, M.R. and MCPHERSON, W.W. 1975. Letter to the editor. Science *192,* 8, 11.

LEACH, G. and SLESSER, M. 1973. Energy equivalents of network inputs to food production processes. Univ. of Strathclyde, Glasgow.

MADDEX, R.L. and BAKKER-ARKEMA, F.W. 1978. Reducing energy requirements for harvesting, drying and storing grain. Energy Fact No. *18,* Extension Bull. *E-1168,* Mich. State Univ., East Lansing.

MAKHIJANI, A. 1975. Energy and Agriculture in the Third World. Ballinger Publishing Co., Cambridge, Mass.

MATTHEWS, J. 1975. Energy consumption in agricultural field work. Span *18* (1) 25–26.

MEIERING, A.G., DAYNARD, T.B., BROWN, R. and OTTEN, L. 1977. Drier performance and energy use in corn drying. Can. Agric. Eng. *19* (1) 49–54.

MILES, J.A. 1975. Energy savings through alternative fuel utilization in desert agriculture. ASAE Paper *75-1004.* Presented at Ann. Meet. Amer. Soc. Agric. Eng., St. Joseph, Michigan

MOREY, R.V., CLOUD, H.A. and LEUSCHEN, W.E. 1976. Practices for the efficient utilization of energy for drying corn. Trans. ASAE *19* (1) 151–155.

NALEWAJA, J.D. 1974. Energy requirements for various weed control practices. Proc. North Central Weed Control Conf. *29,* 18–23.

NEWCOMBE, K. 1975. Energy use in the Hong Kong food system. Agro-Ecosystems *2* (4) 253–276.

ODEND'HAL, S. 1972. Energetics of Indian cattle in their environment. Human Ecol. *1* (1) 3–22.

ODUM, H.T. and ODUM, E.C. 1976. Energy Basis for Man and Nature. McGraw-Hill, New York.

PASSMORE, R. and DURNIN, J.V.G.A. 1955. Human energy expenditures. Physiol. Rev. *35,* 801–840.

PENNER, P. 1974. Summary of transport characteristics for vehicular freight transportation, 1971. CAC Tech. Memo. No. *45,* Cntr. for Adv. Computation, Univ. of Ill., Urbana.

PENNER, P.S. 1975. The dollar, energy and labor impacts of 1971 motor freight transportation. CAC Tech. Memo. No. *44,* Cntr. for Adv. Computation, Univ. of Ill., Urbana.

PIEROTTI, A., KEELER, A.G. and FRITSCH, A.J. 1977. Energy and food. CSPI Energy Series X, Cntr. for Sci. in the Public Interest, Washington, D.C.

PIMENTEL, D. 1974. Energy use in world food production. Environmental Biology Rept. *74–1,* Cornell Univ., Ithaca, New York.

PIMENTEL, D. 1976. World food crisis: Energy and pests. Bull. Entomol. Soc. Amer. *22* (1) 20–26.

PIMENTEL, D., DRITSCHILO, W., KRUMMEL, J. and KUTZMAN, J. 1975. Energy and land constraints in food protein production. Science *190.* 754–761.

PIMENTEL, D. *et al.* 1973. Food production and the energy crisis. Science *182,* 443–449.

PRATT, C.J. 1966. Chemical fertilizers. Sci. Amer. *212* (6) 62–72.

RAO, A.R. and SINGH, I.J. 1977. Bullocks—the mainstay of farm power in India. *In* Energy and Agriculture. W. Lockeretz (Editor). Academic Press, New York.

REGAN, E.J., Jr. 1977. Energy analysis and models of soil formation. *In* Energy Analysis of Models of the United States. H.T. Odum and J. Alexander (Editors). Sys. Ecol. and Energy Analysis Group, Dep. Envir. Eng. Sci., Univ. of Florida, Gainesville.

REVELLE, R. 1976. Energy use in rural India. Science *192,* 969–975.

RICE, R.A. 1972. System energy and future transportation. Tech. Rev. *74,* 31–37.

SCHNEEBERGER, K.C. and BREIMYER, H.F. 1974. Agriculture in an energy hungry world. Ppr. presented at South. Agric. Econ. Assoc. Meet., Memphis.

SCHUFFELEN, A.C. 1975. Energy balance in the use of fertilizers. Span *18* (1) 18–20.

SEBALD, A.V. 1974. Energy intensity of barge and rail freight hauling. CAC Doc. No. *127,* Cntr. for Adv. Computation, Univ. of Ill., Urbana.

SHERFF, J.L. 1975. Energy use and economics in the manufacture of fertilizers. *In* Energy, Agriculture and Waste Management. J. Jewell (Editor). Proc. 1975 Cornell Agric. Waste Mgt. Conf., Ann Arbor Science Publishers, Ann Arbor, Michigan.

SKOLD, M.D. 1977. Farmer adjustments to higher energy prices. The case of pump irrigators. *ERS-663.* Economic Res. Serv.—U.S. Dep. Agric., Washington, D.C.

SLESSER, M. 1973. Energy subsidy as a criterion in food policy planning. J. Sci. Food Agric. *24,* 1193–1207.

SLOGGETT, G. 1977. Energy and U.S. agriculture: Irrigation pumping. 1974. Agric. Econ. Rep. No. *376*, Economic Res. Serv.—U.S. Dep. Agric., Washington, D.C.

SMERDON, E.T. *et al.* 1975. Fuel use estimates for Florida agricultural production. Inst. Food and Agric. Sci., Univ. of Florida, Gainesville.

SPEDDING, C.R.W. 1975. The Biology of Natural Systems. Academic Press, New York.

STANHILL, G. 1974. Energy and agriculture: A national case study. Agro-Ecosystems *1* (3) 205–217.

STANSFIELD, J.R. 1975. Fuel and power in agriculture. Span *18* (1) 23–24.

TUCKER, V.A. 1975. The energetic cost of moving about. Amer. Sci. *63* (4) 413–419.

U.S. BUR. OF CENSUS. 1976. Statistical abstract of the United States. U.S. Govn't. Print. Off., Washington, D.C.

U.S. DEP. AGRIC. (USDA). 1977. Agricultural statistics. U.S. Dept. Agric., U.S. Govn't. Printing Off., Washington, D.C.

WEBB, M. and PEARCE, D. 1975. The economics of energy analysis. Energy Policy *3* (4) 318–331.

WENSINK, R.B., WOLFE, J.W. and KIZER, M.A. 1976. Simulating farm irrigation system energy requirements. *WRRI-44,* Water Resources Research Institute, Oregon State University, Corvallis.

WILLIAMS, D.W. *et al.* 1975. Energy utilization on beef feed lots and dairy farms. *In* Energy, Agriculture and Waste Management. W.J. Jewell (Editor). Ann Arbor Science Publishers, Ann Arbor, Michigan.

7

Evaluation of Alternatives

Increasing population and decreasing resources are presenting an increasingly demanding dilemma for the peoples of this world: How to best use available resources to meet people's needs. Included among the essential requirements for human life is food. The world's population, now above 4 billion, is too large for hunting and gathering or even primitive forms of agriculture to supply its food needs. Presently, industrialized agriculture supplies a sizeable portion of our food and is likely to continue to do so into the forseeable future. In fact, if population continues to increase as predicted, industrialized agriculture will likely supply food to an even larger portion of the world's population in the future.

LESS PRODUCTION AND/OR POPULATION CONTROL

Since current world food production is less than adequate to meet current world population needs, a reduction in food production is actually a nonsolution. It would only be satisfactory if population were also reduced.

It must be realized, however, that this nonsolution would be the result if industrialized agriculture were to be eliminated and replaced with primitive, albeit "energy efficient," forms of agriculture. Persons who should be knowledgeable, including scientists, have, in their criticisms of industrialized agriculture, proposed just this nonsolution.

The world's population will be limited, if not deliberately through birth control or other such means, by starvation, warfare and disease. As our natural nonrenewable resources are consumed, we will find ways to recycle many materials and use renewable energy sources; under such conditions the world's population may decrease to less than the current level.

INTERMEDIATE TECHNOLOGY

An intermediate level of agricultural technology, between primitive and industrialized, has been proposed as one possibility for man's future. Other names for this idea are *appropriate technology* and *mesotechnology*. It is envisioned by some that agriculture might be selectively de-industrialized, keeping some practices and eliminating others, with resulting increased efficiency.

Goldstein (1975) indicated that a low-energy agriculture would emerge due to economic factors. Further, he listed several characteristics of low-energy agricultural systems:

"(1) Greater use of labor-intensive, soil-conserving methods.
"(2) More direct marketing of food between farmer and consumer.
"(3) . . . public acceptance of land application of agricultural and urban organic wastes.
"(4) More people able to earn sufficient income while living in rural areas.
"(5) A recognition of the interrelationship between farms and cities."

Many opportunities exist to observe and analyze intermediate technology in its infinite variety, as segments of agriculture around the world transform from primitive to semi-industrial. Also, energetic analyses have been made for several such systems existing in industrialized economies. Johnson *et al.* (1977) studied the Amish, who live in communities in Pennsylvania, Indiana and other states. The Amish farm uses methods which are labor intensive and which are only partially mechanized in comparison with their non-Amish neighbors. Some Amish do not use electric power, nor any internal combustion engines except stationary ones. They rely heavily upon horses as their main source of power. Consumption of fertilizers and pesticides is considerably less. Comparisons with non-Amish indicated that the Amish consume considerably less energy, both as agricultural inputs and particularly to support their lifestyle. Their yields are less, but the energy ratios of their products are larger. Thus, a widespread shift from modern industrialized agriculture to Amish-type farming would result in a more energy efficient production system, but would also result in less food produced.

Burnett (1978) analyzed three examples of intermediate technology which were entries in a *Mother Earth* food self-sufficiency contest. The three, when compared with United States agriculture and society, evidenced low consumption of energy both for agriculture and for lifestyle support, and energy ratios both above and below the average for United States agriculture.

The energetics of organic (low energy) versus conventional (high energy) wheat production were analyzed by Berardi (1977). The organic farmers had 31% less energy inputs to wheat production, 22% less production and 12% higher energy productivity than the conventional farmers.

Lockeretz *et al.* (1975) compared a group of farms that did not use inorganic fertilizers and pesticides with a group which did and found the organic farms had lower energy inputs, somewhat lower yields for most crops, and the same net profit.

Intermediate technology seems to be best suited for areas where soils are less fertile, terrain is less suited for mechanization, fields are small and irregular in shape, and livestock and diversification are features of the total agricultural system.

INCREASED ENERGY PRODUCTIVITY

One desirable and reasonable goal is for agricultural systems to have increased energy productivity. Increased energy productivity is indicative of greater production per unit of input energy, and therefore of more efficient usage of energy.

Figure 7.1 shows three ways by which increased energy productivity can be achieved. Area 1 shows that increased energy productivity can be achieved even when yields are decreased if energy inputs are decreased more than are yields (possibly resulting from intermediate technology). Possible disadvantages of this approach are that less food will be produced and that profitability may be reduced. Area 2 indicates perhaps the most desirable combination of energy inputs and yields for increased energy productivity, increased yield and decreased energy consumption. Even in this situation, however, it is possible that decreased energy consumption may incur sufficiently increased costs to reduce profitability. Area 3 indicates increased energy productivity through increased yields greater than increased energy consumption (possibly resulting from higher intensity agriculture). Again, energy productivity, although increased, does not guarantee economic feasibility.

It should therefore be recognized that changes in agricultural systems which affect yield or productivity also affect the energetics of the system, even though the energy inputs may not be changed in quantity, or

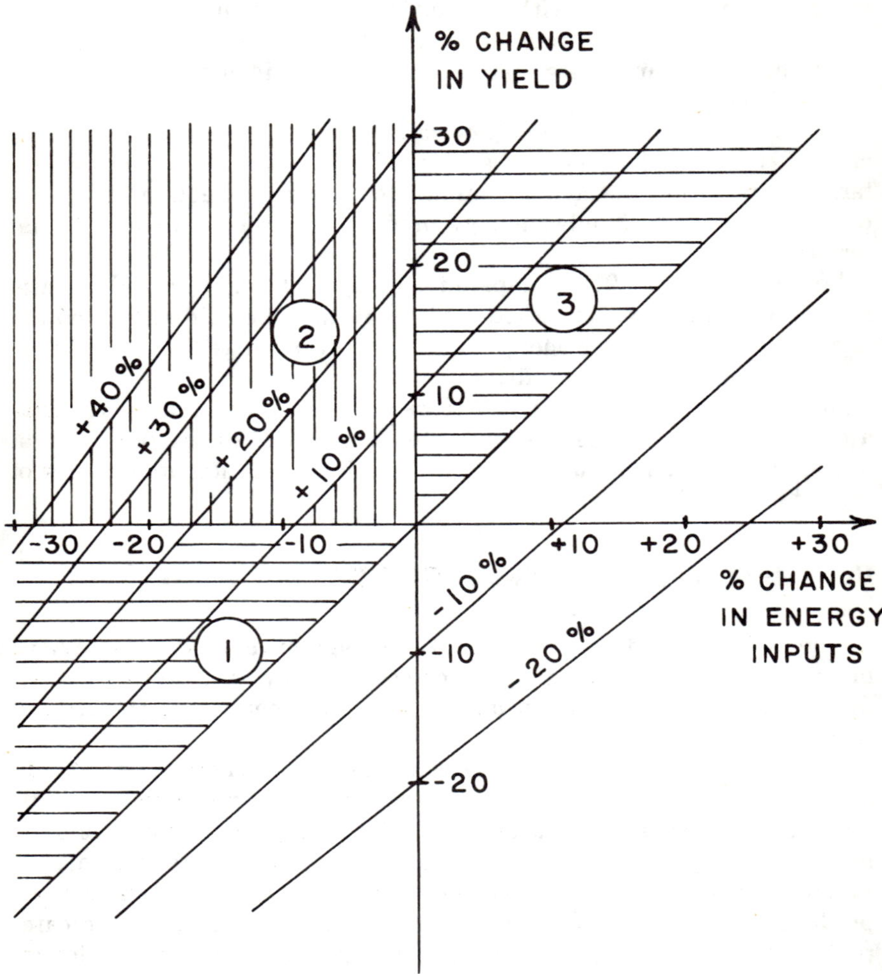

FIG. 7.1. EFFECTS OF CHANGES IN YIELD AND ENERGY INPUTS ON ENERGY PRODUCTIVITY

conceivably even in source. Any advance in agricultural technology, therefore, has implications concerning the energetics of agricultural systems.

One method of analyzing the energy productivity of an agricultural production practice is through examination of the response of yield to energy input. Such curves may be drawn for the summation of all inputs or for a single energy-consuming input. Energy productivity is by definition, the quotient of yield and energy inputs. For a curve showing yield response to a single energy-requiring input, partial energy productivity is the slope of a line drawn from the point of intersection of the curve with the ordinate to the point of the curve corresponding to the level of input being examined. A hypothetical curve (Fig. 7.2) shows greatest slope and, therefore, greatest partial energy productivity at the point of the curve tangent to the drawn line. It may or may not be desirable to apply the level of energy-requiring input which results in maximum partial energy productivity—the point of maximum profits will not in general coincide—but when such a curve is known, the effects on energy productivity and yield of a change in the level of input energy can be determined.

Any method by which increases in yield or productivity can be obtained has implication for the energetics of agricultural systems. Some of the more important ones will now be examined.

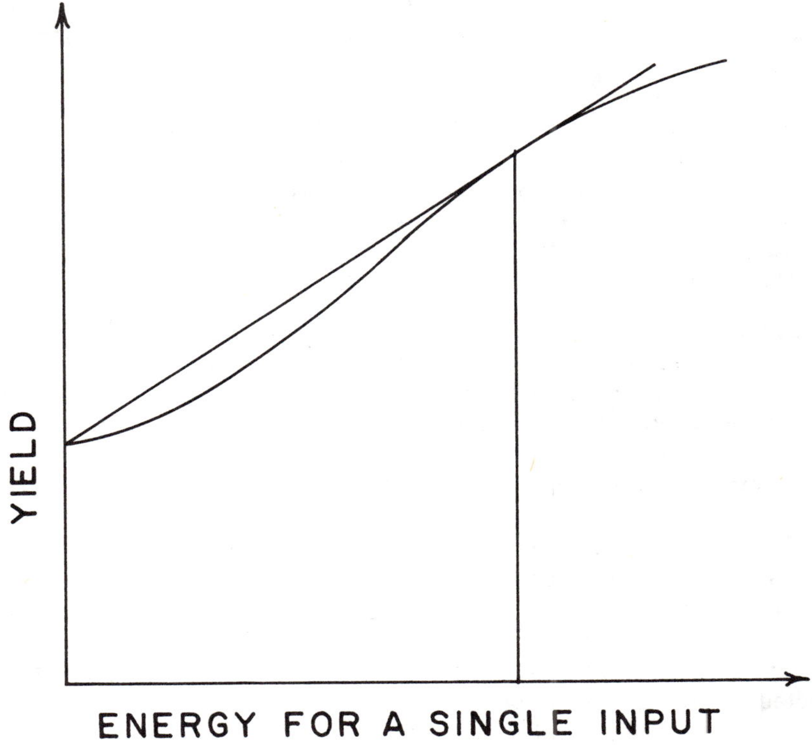

FIG. 7.2. HYPOTHETICAL CURVE OF YIELD VERSUS INPUT ENERGY FOR A SINGLE INPUT SHOWING A UNIQUE POINT OF MAXIMUM PARTIAL ENERGY PRODUCTIVITY

Reduction of Losses.—One area deserving much attention and effort is that of reducing losses of agricultural products. Reduction of losses increases the amounts of agricultural product available for consumption just as does increased production, and often with less additional energy inputs. In industrialized agricultural systems, losses occur during production, and particularly at harvesting, and throughout the food chain, even after food preparation. In tropical countries, where temperatures are higher and pests are prevalent, preservation of foods generally is more difficult and losses are greater. Smerdon (1977) reviewed available information concerning the magnitude of agricultural product losses. In India it was reported that as much as 50% of the food raised is lost to pests and parasites; rodents alone are thought to destroy 25% in the field or in storage. Worldwide losses may amount to 30%, and 10% of stored food may be lost on a worldwide basis.

Pimentel (1976) reviewed losses and estimated United States preharvest crop losses due to insect, pathogen and weed pests at about 33%, and postharvest losses at 6% for total losses of 39%. Further, worldwide losses were estimated at 35% preharvest and 13% postharvest for total losses of 48%.

If food losses could be reduced by half, say from 30 to 15% with an increase in energy inputs of approximately 5%, energy productivity would be increased by 15.6% (Fig. 7.1).

Kling's (1943) estimates of total losses varied with product. Eggs and sugar were lowest at 4%. At the opposite end of the loss spectrum he found dry legumes and nuts at 50% and leafy, green and yellow vegetables at 43%.

Use of Fertilizers.—Additional fertilizer typically results in increased yields with the yield increases becoming less at higher fertilization rates and perhaps peaking and then declining with additional increases. Energy productivity, therefore, is greatest at lowest fertilizer application rates and continually declines with additional increases in fertilizer application rates, through and beyond the point of maximum yields (Fig. 7.3). Under conditions of limited fertilizer availability, a farmer should apply the fertilizer he/she has for a crop uniformly to all the land planted to that crop, rather than applying it at a higher rate to some of the land in order to achieve higher yields and at a lower rate to the remainder. Total yields are maximized with uniform application. Similarly, a more nearly uniform worldwide distribution of total fertilizer would be expected to increase total worldwide agricultural production.

Soil testing is essential to measure the nutrients available and enable recommendations for fertilizer application rates. Once every three years

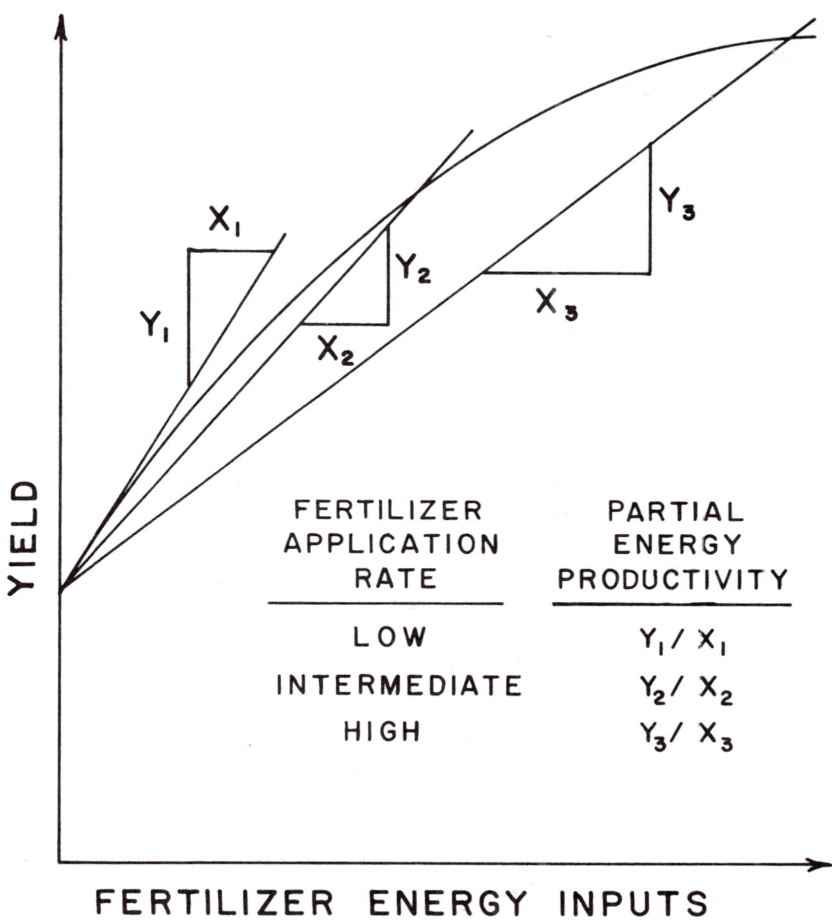

FIG. 7.3. EFFECT OF FERTILIZER APPLICATION RATE ON PARTIAL ENERGY PRODUCTIVITY

is a recommended frequency for average conditions. Following the recommendations based on soil testing enables more efficient use of fertilizer and of the energy required to provide the fertilizer. Farmers tend to overfertilize high value cash crops such as fruits, vegetables and tobacco and to underfertilize field crops, forages and pasture. Better distribution of fertilizer among over- and under-fertilized crops could lead to higher yields and increased energy productivity.

Liming to raise the pH of the soil can affect efficiency of fertilizer nutrient usage since availability of phosphorus to plants, nodulation and subsequent nitrogen fixation of legumes is affected by pH.

Placement of fertilizer with respect to the plant can affect fertilizer efficiency. Proper banding of fertilizer alongside the plants in a row can lead to greater yields than general broadcast application of the same amount.

Losses of nutrients by leaching through the root zone due to too great quantities of rainfall or irrigation is a problem, particularly in sandy soils. Multiple applications throughout the growing season, application of water soluble nutrients through the irrigation system, avoidance of excessive irrigation, use of slow release materials and plastic mulches are all methods by which leaching of nutrients may be reduced or avoided.

The use of a nitrogen-fixing green manure cover-crop plowed under in order to provide nutrients will normally increase energy productivity but can have varying effects on overall yields. Since legumes can fix 170–450 $kg \cdot ha^{-1} \cdot yr^{-1}$ of nitrogen requiring perhaps 12–30,000 $MJ \cdot ha^{-1} \cdot yr^{-1}$ to replace with chemical fertilizers, energy productivity will likely be increased substantially. However, if the growing of the cover crop eliminates another crop, the long-term yield of the other crop will be reduced. If the cover crop can be grown during the off-season for the other crop, the practice would seem to be energetically sensible, not result in yield decreases, and perhaps be economically feasible.

Use of Pesticides.—Many pesticides are used to increase production or reduce losses. Other pesticides are used to reduce threats to human health, decrease discomfort, etc.

Whether or not use of pesticides results in increased production or increased energy productivity depends on many factors, not all of which are controllable or likely known. The effectiveness of a herbicide, for example, in reducing a weed population, may depend on temperature, manner of application, soil moisture conditions and many other factors.

The response of yield to quantity of a pesticide used may be hypothesized to be an increasing function. However, the shape of the response curve is likely different from response curves for fertilizers. A hypothetical yield response to pesticide application is shown in Fig. 7.4. At low rates the pesticide does not sufficiently reduce pest populations to allow yields to increase very much above the pest-reduced levels. At high rates, the pest population is greatly reduced and yields are maximized. Therefore, a sigmoid-shaped curve results. Partial energy productivity is maximized at the specific pesticide application rate shown. At rates either higher or lower than the optimum application rate the slope of the line drawn from yield at zero rate to yield at the application rate will be less and the partial energy productivity will be less.

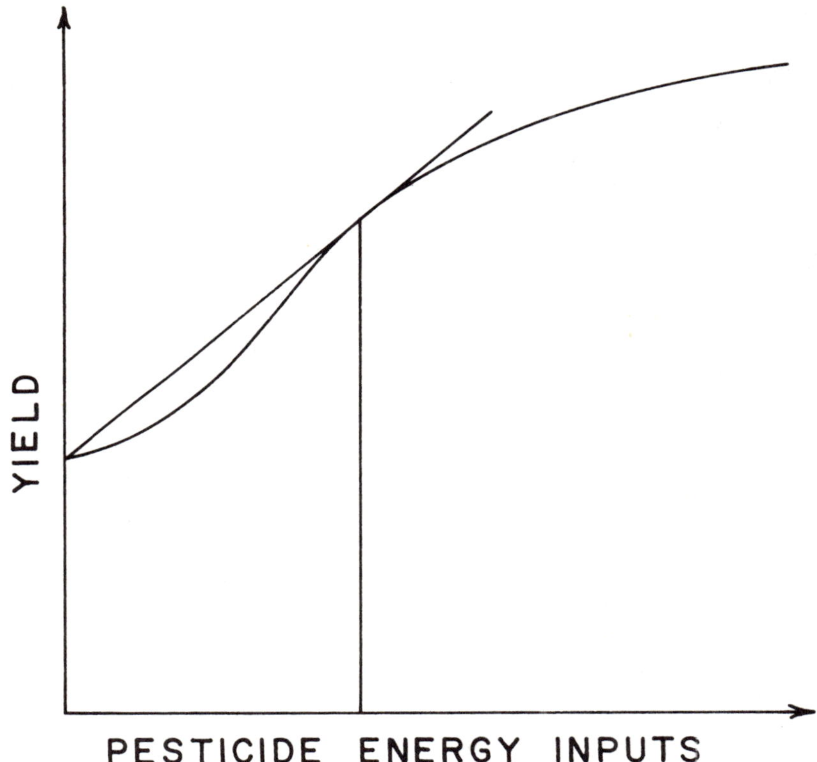

FIG. 7.4 EFFECT OF PESTICIDE APPLICATION RATE ON PARTIAL ENERGY PRODUCTIVITY

Use of Mechanization.—Mechanization in agricultural production has had two major results: increases in productivity (termed by Giles 1975 as yield mechanization) and reductions in labor intensity. In industrialized agriculture both effects have been important. In subsistence agriculture, where labor is abundant most times during the year, a reduction in labor intensity would usually be undesirable, as it will increase unemployment. However, mechanization may be desirable in subsistence agriculture if labor becomes limiting; if this occurs it will usually be at planting or harvesting.

The authors will attempt here to separate these two results of mechanization and look only at the effects of mechanization on productivity. One of the first to recognize and attempt to quantify this phenomena was Giles (1967). Figure 7.5 shows his first attempt, which was also presented by Odum (1971). Giles inferred from such data that a

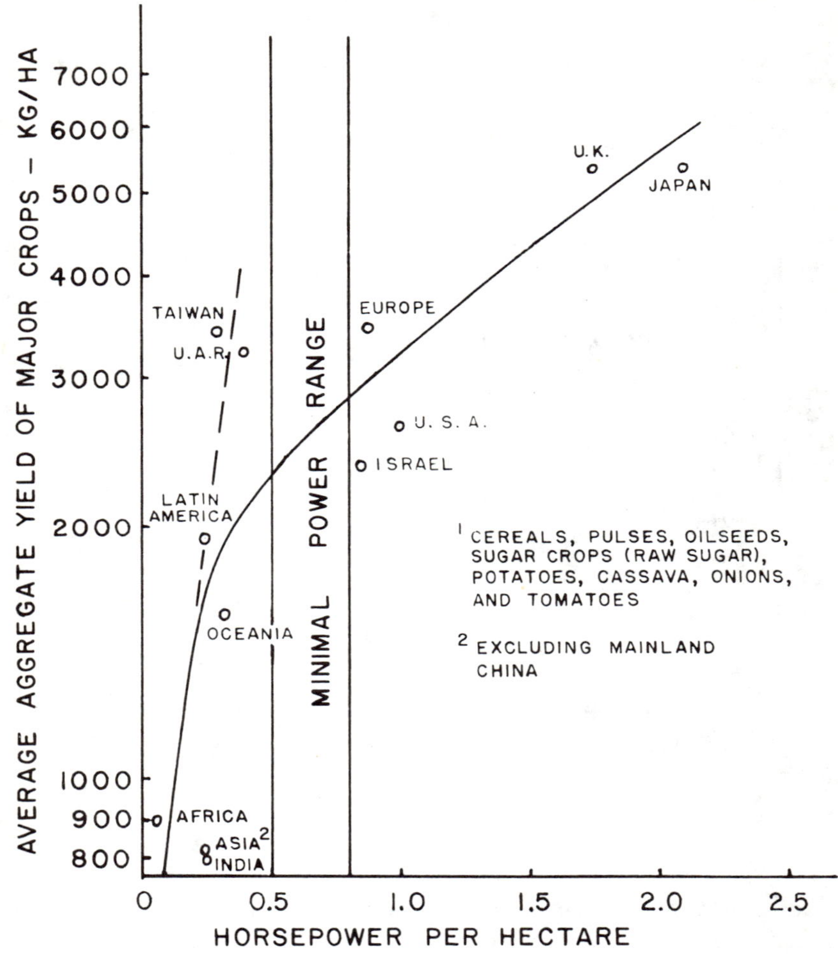

From Giles (1967)

FIG. 7.5. RELATIONSHIP BETWEEN YIELDS IN KG/HA AND POWER IN HP/HA FOR MAJOR FOOD CROPS

minimum power availability of from 0.5 to 0.8 hp $\cdot$ ha^{-1} is necessary for respectable yields. Note that this power is not applied continuously but only periodically when needed for specific operations. On such a basis a Taiwanese farmer with 5 ha should have a 3 or 4 hp walking tractor to produce respectable yields.

Two similar graphs, but both including all inputs rather than only mechanization, were presented by Odum (1971B) (Fig. 7.6) and Slesser

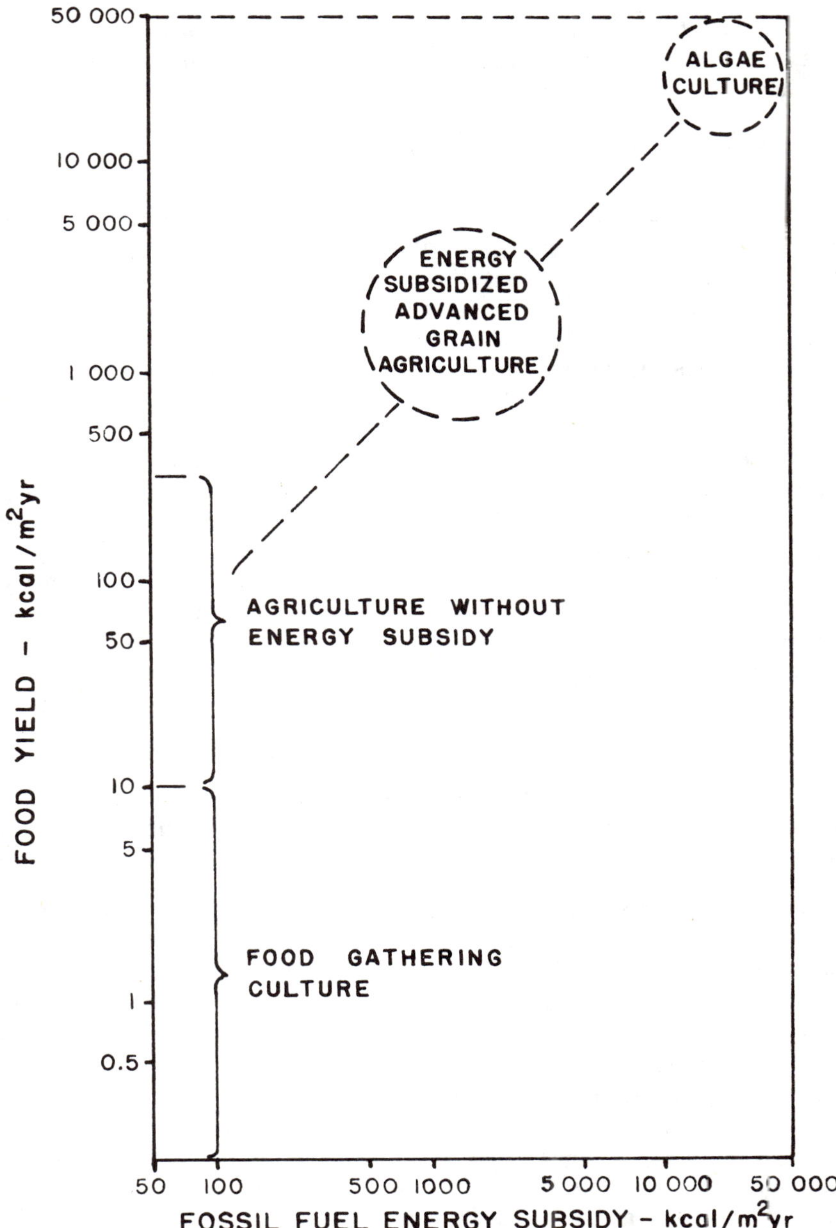

From Environment, Power, and Society, Odum (1971). Copyright © 1971. Used with permission of John Wiley & Sons, Inc.

FIG. 7.6. NET FOOD YIELDS TO MAN AS A FUNCTION OF THE SUBSIDY OF FOSSIL-FUEL INDUSTRY

(1973). The latter was a plotting on logarithmic scales of protein yield of a large number of agricultural systems as a function of energy subsidy (mechanical inputs, transportation, nutrients, fertilizers, pesticides, irrigation, seed, etc.). Slesser gave the regression equation P = 1.415 e.s. 0.817 where P is $kg \cdot ha^{-1} \cdot yr^{-1}$ of protein produced and e.s. represents the energy subsidy in 10^3 $kcal \cdot ha^{-1} \cdot yr^{-1}$.

Singh and Chancellor (1975) studied the effects of mechanization on Indian agriculture. As mechanization levels increased, energy inputs per unit of crop output increased, labor inputs per unit of crop output decreased, and production costs per unit of crop output decreased.

Williams and Chancellor (1975) investigated the effects on production due to changes in the inputs of fertilizer, irrigation and mechanization (separated into three inputs: tillage and planting power, harvest capacity and production energy) on nine California-grown crops. Each of these inputs was considered as limiting maximum production if input quantities were reduced below optimum levels. Reductions in the levels of these inputs from optimum to half of optimum resulted in production decreases of 49% for irrigation (California agriculture is heavily dependent upon irrigation), 29% for fertilizer, 26% for production fuel energy, 21% for harvest capacity, and 15% for tillage and planting power.

From the available data it is hypothesized that response curves of yield to level of mechanization as indicated by power availability are shaped as shown in Fig. 7.7. Curves for yield as a function of mechanization energy inputs would also have the same shape. Therefore, the highest partial energy productivity is achieved at the point of minimum mechanization energy inputs, and increasing mechanization energy inputs increases yields at a decreasing rate. Implications are similar to those for fertilizer.

Use of Irrigation.—Irrigation may either supplement natural rainfall when necessary or provide essentially all the crop's water requirements. The effects, therefore, of irrigation application on yields are quite varied, depending greatly upon rainfall where irrigation is supplemental. Where irrigation is the only source of water, crop yields are more predictably based upon irrigation amounts and timing.

A hypothetical yield response to irrigation energy inputs is shown in Fig. 7.8. It shows, in general, increasing yields at a decreasing rate and sharply decreasing yields with too much irrigation (soil moisture beyond field capacity). The inflection at the beginning of the curve is indicative of the indirect energy inputs of the irrigation system necessary even if the system is unused. The curve should be shifted downward so that it intersects zero yield for those systems in which irrigation is essential for crop production.

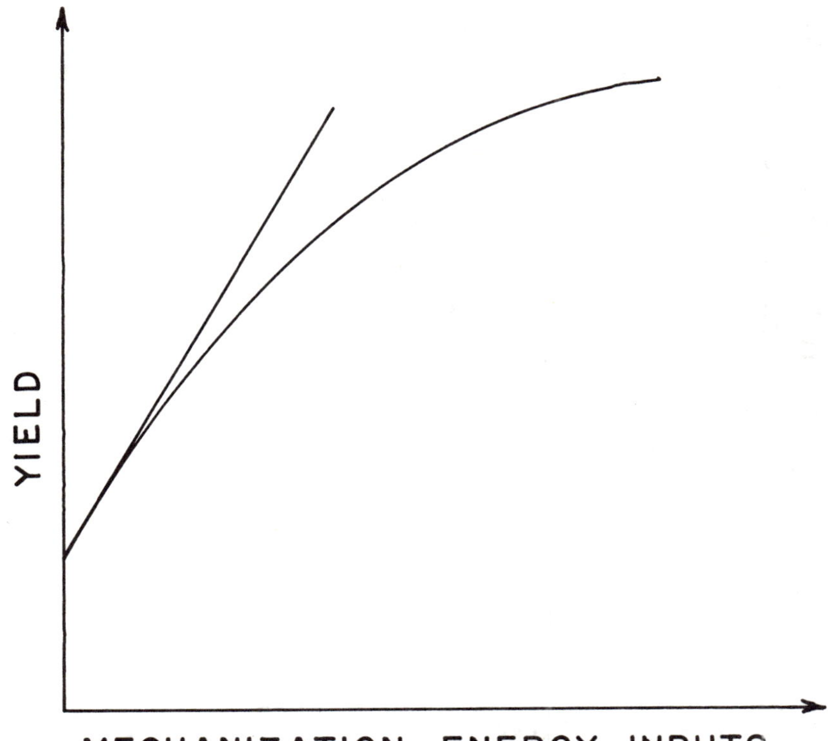

FIG. 7.7. EFFECT OF LEVEL OF MECHANIZATION ON PARTIAL ENERGY PRODUCTIVITY

Maximum partial energy productivity is achieved at the level of irrigation energy inputs indicated in Fig. 7.8 by the line tangent to the yield curve.

Scientific Advances.—The future course of technology, development of improved methods and scientific breakthroughs are impossible to predict. However, the total application of new knowledge is steady and resultant productivity does seem to generally increase. The U.S. Department of Agriculture's (USDA 1977) agricultural productivity index, relating all agricultural outputs to all inputs, has increased over the past 20 years at about 1.3% annually.

Berg (1978) has suggested that a crisis attracts attention and talent and resources, which results in scientific advances. Certainly the energy crisis and resultant research and development efforts will have effects. Increases in energy productivity due to scientific advances may be

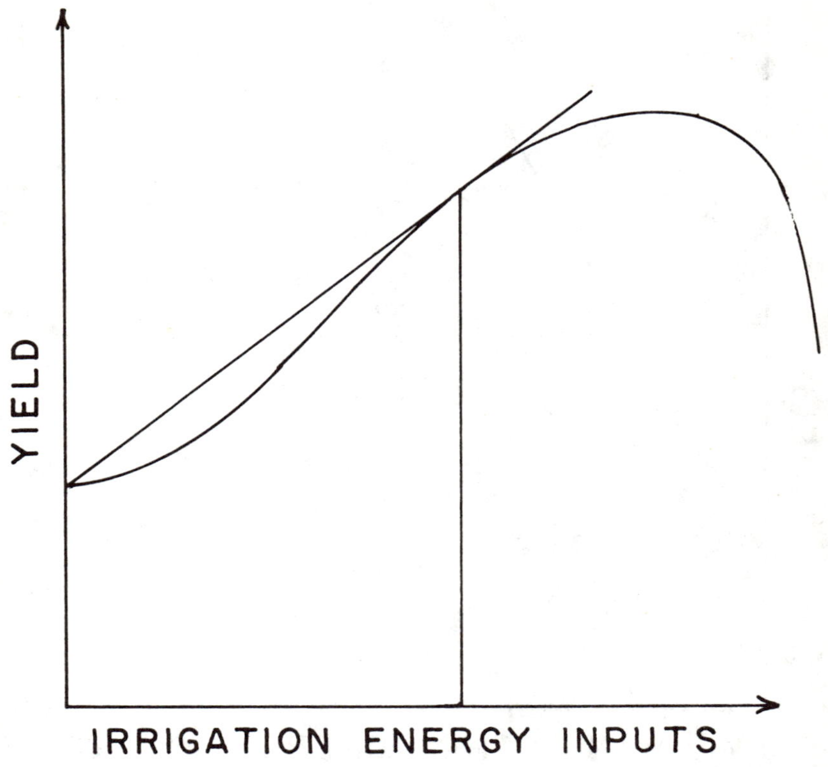

FIG. 7.8. EFFECT OF IRRIGATION ENERGY INPUTS ON PARTIAL ENERGY PRODUCTIVITY

expected to occur both from the whole spectrum of advances resulting in improved productivity and from those focusing specifically on energy.

Multiple Cropping.—Multiple cropping is the growing of more than one crop annually. It is widely used where the growing season is sufficiently long or where sufficiently mild winters allow crops to be grown a large portion of the year. Multiple cropping may result in increased energy productivity. If primary tillage is performed only once per year and two crops are grown during the year, such as corn followed by soybeans, the energy required for primary tillage may be apportioned between the two crops rather than being required for each. Thus, the energy inputs for each crop are less than they would otherwise have been if only one crop were grown. Further, soil nutrients will likely be better utilized and less nutrient losses occur due to leaching, so that total fertilizer applications may be less than if the crops were grown in separate years.

Timeliness.—Much of the increased yields associated with increased power levels of draft animals and mechanization may be attributed to better timeliness of crop operations. An American Society of Agricultural Engineers (ASAE) committee on machinery management (1978) has defined timeliness as the "ability to perform an activity at such a time that quality and quantity of product are optimized." Planting and harvesting dates are among the most critical in affecting yields. Timing of pest control operations may be critical. Some operations in livestock enterprises may affect the quality and quantity of products and therefore have applicability to the concept of timeliness.

As an example of timeliness, Chancellor and Cervinka (1974) examined the effects of the dates of planting and harvesting operations for California rice production. Figure 7.9 shows the effect of harvest date on crop value and Fig. 7.10 shows the combined effect of planting date and harvesting date. Performing operations just a few days from the optimum dates reduced yields 5–10%.

Energy productivity may be substantially increased by improving timeliness. Yields are increased with better timeliness. Energy inputs may also be increased but, if so, usually only for the indirect energy inputs to provide for increased machinery capacity. The use of otherwise oversized and overpowered equipment may be justifiable due to increased yields obtained through improved timeliness.

Substitution of Inputs.—An entrepreneur combines quantities of various inputs to provide a product or service at least total cost. When the price of one input increases relative to the others, one or more lower-priced inputs may be partially or totally substituted for the one whose price has most increased. A competitive economy assures that this adjustment of inputs is continual.

When the price of energy increases relative to other inputs which can be wholly or partially substituted for energy, efforts will be made to replace some of the energy inputs by materials, labor, land, etc. Also, when one form of energy increases in price relative to others, efforts will be made to make substitutions of lower cost forms of energy.

Energy has risen in price relative to many other inputs, largely due to the sharp price increases of oil in 1973–74 and their subsequent effects on the prices of other sources of energy. The expected result of substitutions for higher priced energy is that energy productivity of agricultural systems will increase. This increase will be at the expense of greater consumption of other inputs. Further, it indicates nothing about the

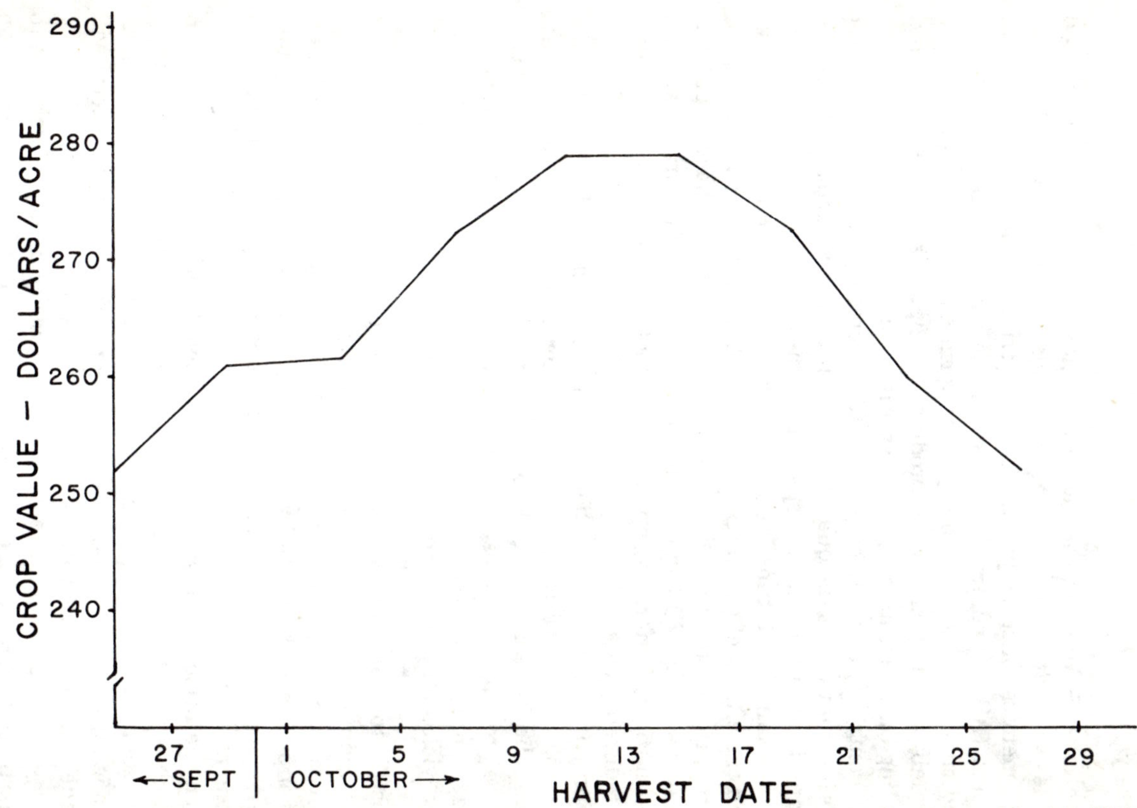

From Chancellor and Cervinka (1974)

FIG. 7.9 EFFECT OF HARVESTING DATE ON CROP VALUE FOR "CALROSE" RICE PLANTED APPROXIMATELY APRIL 27 IN RICHVALE, CALIFORNIA AREA

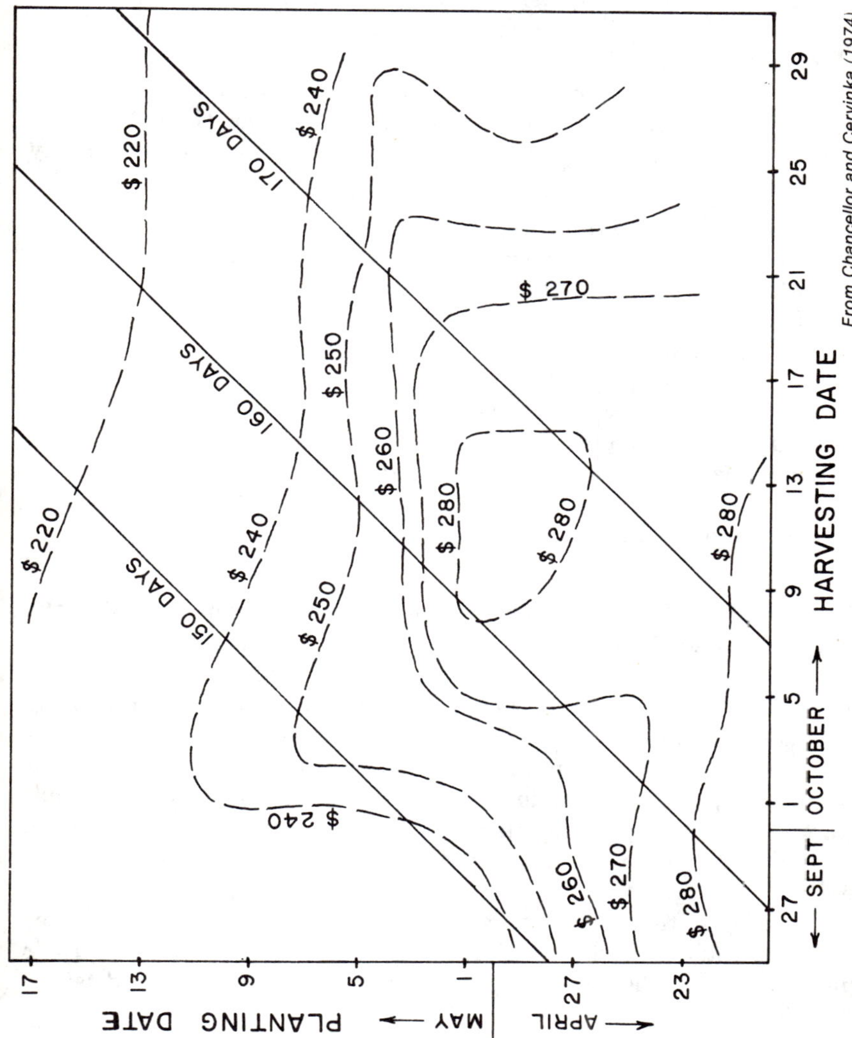

From Chancellor and Cervinka (1974)

FIG. 7.10 EFFECTS OF PLANTING AND HARVESTING DATES ON CROP VALUE FOR "CALROSE" RICE IN CALIFORNIA

total yields and whether they may increase or decrease as a result of such substitutions.

Additional decreases, in the amounts of indirect energy inputs, may be expected as less energy-intensive inputs are substituted for ones which are more energy intensive. If such a substitution were to have no effect on yield, energy productivity would be increased.

Relating materials conservation to energy conservation, Purcell (1977) gave a listing of means to conserve resources. Any of these may also reduce energy inputs and may therefore result in increased energy productivity:

"(1) Make/use less products
"(2) Make more durable or reusable products
"(3) Make more energy efficient products
"(4) Use products longer
"(5) Make/use lighter products
—using less materials
—using lighter weight material
"(6) Substitute less energy-intensive materials in products
"(7) Recycle waste products
—to energy
—to raw materials"

Conservation of Inputs.—The potential for conserving energy in the food system is significant, even though the food system consumes only about 17% of our total energy consumption, and even though the potential for savings in areas such as transportation and space heating and air conditioning is likely much greater. The Federal Energy Administration (FEA 1975) has produced detailed estimates of energy consumption in agricultural production. Opportunities for conservation are in part dependent upon the magnitude of consumption; therefore,the following discussion will focus on those operations consuming more energy.

It is implicit in the consideration of most energy conservation practices that yields would be either minimally decreased or unaffected by their implementation. Otherwise energy productivity will be decreased rather than increased.

Fertilization.—Fertilizers likely require more energy than any other agricultural input. FEA (1975) estimated 0.655×10^{12} MJ for the United States in 1974. Methods which may result in less fertilizer consumption without corresponding reductions in production include:

(1) Increased soil testing and following of recommendations.
(2) Better application methods to place fertilizer where needed by the plant.
(3) Better timing of applications so that fertilizer will be available when needed by the plant.
(4) Application of fertilizer nutrients through irrigation systems.
(5) Use of plastic mulch to reduce leaching losses of fertilizer nutrients by excess rainfall.
(6) Multiple cropping to use residual fertilizer nutrients from prior crop before leaching losses occur.
(7) Substitution of manures, cover crops and legumes.

Field Operations.—FEA (1975) estimated 1974 United States direct energy consumption for all field operations as totaling 0.412×10^{12} MJ. Field operations included preplant (plowing, discing, harrowing, etc.), plant, cultivate, fertilizer and pesticide application, and harvest. Most field operations involve tractors, although some are performed with self-propelled equipment, particularly self-propelled harvesters. Practices which may conserve energy in field operations include:

(1) Operate tractor at most efficient speed and load, popularly expressed as "Gear up and throttle down."
(2) Keep internal combustion engines optimally tuned and maintained.
(3) Reduce unnecessary idling.
(4) Combine operations whenever possible, such as in forming a seedbed, applying herbicide and fertilizer and planting, all in a single pass through the field.
(5) Optimum ballasting. Too much tractor weight results in increased fuel consumption and soil compaction; too little results in increased wheel slippage and tire wear.
(6) Minimum tillage or reduced tillage, in which crops are planted in soil without the usual plowing, discing, etc. Herbicide usage is increased.
(7) Shift from gasoline to diesel.
(8) Matching of tractor size to load.
(9) Use of plastic mulch on horticultural crops thereby reducing fertilizer application and cultivation frequency.

Irrigation.—Irrigation is likely the third largest energy-using procedure in United States agricultural production. FEA (1975) estimated 0.275×10^{12} MJ in direct energy consumption for 1974. Practices which may result in less energy consumption for irrigation include:

(1) Application of no more water than needed for the crop.
(2) Shifting to less energy intensive application methods where feasible.
(3) Increased pump and engine efficiency through proper design, selection and maintenance.
(4) Reduce evaporation losses, particularly with sprinkler irrigation, by nighttime application.
(5) Increase application efficiency so that total amount of water applied is reduced.
(6) Reduction of pressure at sprinkler heads to minimum feasible pressure.

Crop Drying.—FEA (1975) estimated direct energy consumption for United States 1974 crop drying was 0.111×10^{12} MJ. Much of the United States corn crop and several other crops including tobacco, soybeans, rice and peanuts are dried with energy supplied by fossil fuels, typically in the form of LP or natural gas. Practices which may reduce the energy required for crop drying include:

(1) Delaying harvest to allow as much field drying as possible.
(2) Convert to low-temperature drying systems where possible.
(3) Use of unheated air for drying whenever feasible.
(4) Use of solar energy for heating drying air.
(5) Avoid overdrying.
(6) Use of dryeration, the slow cooling of grain which has been removed hot from a dryer and held for several hours with no air flow.
(7) Use of high moisture grain for feeding hogs.
(8) Choosing varieties having lower moisture content at harvest.
(9) Preservation of grain with organic acids rather than drying.

Pesticides.—FEA (1975) estimated that 0.101×10^{12} MJ were required to manufacture pesticides used in 1974 in United States agriculture. Conservation practices related to the use of pesticides are:

(1) Integrated Pest Management (IPM) programs; IPM requires a higher level of management; pesticides are applied on a symptomatic rather than a prescheduled basis.
(2) Pest resistant varieties and breeds.
(3) Biological control.
(4) Better application methods to reduce application rates and drift, such as electrostatic spraying.

Transportation.—Transportation, in the broadest sense as related to agriculture, includes the movement of input materials from their points of manufacture to use, movement of management and labor, and movement of products from the farm. FEA (1975) estimated total direct energy use for trucks and autos for 1974 United States agriculture at 0.334×10^{12} MJ; this does not include much of the energy required for transporting inputs to the farm or products from the farm.

Measures to conserve energy in agricultural transportation include those applicable to all transportation. Some of these are:

(1) Selection and purchase of more efficient vehicles.
(2) Proper maintenance.
(3) Good driving habits.
(4) Substitution of communication for personal transporation.
(5) Planning ahead to combine trips.

Frost Protection.—The burning of fossil fuels and the forced movement of air with wind machines and helicopters are efforts to moderate the damaging effects of low temperatures on crops grown in fields and groves. FEA (1975) estimated 1974 United States agricultural energy use for frost protection at 0.045×10^{12} MJ. Most of the energy used in frost protection is for horticultural crops. Methods for conserving energy used for frost protection include:

(1) Use of sprinkler irrigation rather than heating or moving air.
(2) Shifting from gasoline engines to electric motors for wind machines.
(3) Delaying of blossoming of deciduous fruit crops by evaporative cooling with sprinkler irrigation.
(4) Heating to only protect citrus trees rather than fruit.
(5) Use of better temperature prediction methods, especially satellite forecasting, to reduce protection period.

BIBLIOGRAPHY

AMER. SOC. AGRIC. ENGIN. (ASAE). 1978. Uniform terminology for agricultural machinery management, ASAE Standard: ASAE *S322. In* 1978–79 Agricultural Engineers Yearbook. Amer. Soc. Agric. Eng., St. Joseph, Michigan.

BERARDI, G.M. 1977. An energy and economic analysis of conventional and organic wheat farming. *In* Food, Fertilizer and Agricultural Residues, Proc. of

1977 Cornell Agric. Waste Mgt. Conf. Raymond C. Loehr, (Editor). Ann Arbor Sci. Publishers, Ann Arbor, Michigan.

BERG, C.A. 1978. Process innovation and changes in industrial energy use. Science *199,* 608–614.

BUFFINGTON, J.D. and ZAR, J.H. 1976. Realistic and unrealistic energy conservation potential in agriculture. *In* Agriculture and Energy. W. Lockeretz (Editor). Academic Press, New York.

BURNETT, M.S. 1978. Energy analysis of intermediate technology agricultural systems. M.S. Thesis, Univ. of Fla., Gainesville.

CHANCELLOR, W.J. and CERVINKA, V. 1974. Timeliness coefficients for rice and factors affecting their value. Trans. ASAE *17* (5) 841–844.

COBLE, C.G. and LEPORI, W.A. 1975. Energy Consumption, Conservation, and Projected Needs for Texas Agriculture. Rept. S/D-12. Dep. of Agric. Eng., Texas A & M Univ., College Station.

COMMONER, B., GERTLER, M., KLEPPER, R. and LOCKERETZ, W. 1974. The Effect of Recent Price Increases on Field Crop Production Costs. *CBNS-AE-1,* Cntr. for Biol. of Nat. Sys., Washington Univ., St. Louis, Mo.

COUNC. AGRIC. SCI. TECHNOL. (CAST). 1975. Potential for Energy Conservation in Agricultural Production. Rept. No. *40,* Counc. for Agric. Sci. and Technol., Dep. of Agronomy, Iowa State Univ., Ames.

EDMINSTER, T.W. 1975. Weeds, Energy and You. Address at 15th Meet. of Weed Sci. Soc. of Amer., Washington D.C.

ECONOMIC RES. SERV. (ERS). 1974. The U.S. Food and Fiber Sector: Energy Use and Outlook. Economic Res. Serv.—U.S. Dep. Agric., Washington, D.C.

FED. ENERGY ADMIN. (FEA). 1975. Energy and U.S. Agriculture: 1974 Data Base. Fed. Energy Adm.—U.S. Dep. Agric., Washington, D.C.

FLUCK, R.C. *et al.* 1977. Florida Agricultural Energy Conservation Measures. Rept. submitted to State Energy Office and Fla. Dep. of Agric. for inclusion in Fla. Energy Conservation Plan, Agric. Eng. Dep., Univ. of Fla., Gainesville.

GILES, G.W. 1967. Agricultural power and equipment. *In* The World Food Problem Vol. III, Rept. of President's Sci. Adv. Comm., White House, Washington, D.C.

GILES, G.W. 1975. The reorientation of agricultural mechanization for the developing countries, Part I, policies and attitudes for action programs. Agric. Mechanization in Asia *6* (2) 1–16.

GOLDSTEIN, J. 1975. The emerging economic base for low-energy agriculture. *In* Energy, Agriculture and Waste Management, Proc. 1975 Cornell Agric. Waste Mgt. Conf. William J. Jewell (Editor). Ann Arbor Science Publishing Co., Ann Arbor, Michigan.

GREEN, M.B. and McCULLOUCH, A. 1976. Energy considerations in the use of herbicides. J. Sci. Food Agric. *27,* 95–100.

JOHNSON, W.A., STOLTZFUS, V. and CRAUMER, P. 1977. Energy conservation in Amish agriculture. Science *198,* 373–378.

JONES. D.P. 1975. The energy relations of pesticides. Span *18* (1) 20–22.

KLING, W. 1943. Food waste in distribution and use. J. Farm Econ. *25,* 848–859.

LEACH, G. and SLESSER, M. 1973. Energy Equivalents of Network Inputs to Food Production Processes. Univ. of Strathclyde, Glasgow.

LOCKERETZ, W. *et al.* 1975. A comparison of production economic returns, and energy intensiveness of corn belt farms that do and do not use inorganic fertilizers and pesticides. *CBNS-AE-4,* Cntr. for Biol. of Nat. Sys., Washington Univ., St. Louis, Mo.

ODUM, H.T. 1971. Environment, Power and Society. Wiley-Interscience, New York.

PIMENTEL, D. 1976. World food crisis: Energy and pests. Bull. Entomol. Soc. Amer. *22* (1) 20–26.

PIMENTEL, D. *et al.* 1973. Food production and the energy crisis. Science *182,* 443–449.

PURCELL, A.H. 1977. Materials and energy conservation: What lies ahead? *In* Energy Technology V, Government Institutes, Inc., Washington, D.C.

SINGH, G. and CHANCELLOR, W. 1975. Energy inputs and agricultural production under various regimes of mechanization in northern India. Trans. ASAE *18* (2) 252–259.

SLESSER, M. 1973. Energy subsidy as a criterion in food policy planning. J. Sci. Food Agric. *24,* 1193–1207.

SMERDON, E.T. 1974. Energy conservation practices in irrigated agriculture. Pres. at Sprinkler Irrigation Assoc. Ann. Tech. Conf., Feb. 24, Denver.

SMERDON, E.T. 1977. The role of energy in food production—a perspective. *In* Energy Use Management, Proc. Int. Conf. on Energy Use Mngt., Oct. 24–28. Tucson, Arizona.

SPLINTER, W.E. 1977. Energy options for the irrigator. Irrigation Age 50–51, 54–56.

U.S. DEP. AGRIC. (USDA). 1965. Losses in agriculture. Agric. Hdbk. No. *291.* Agric. Res. Serv.—U.S. Dep. Agric., Washington, D.C.

USDA. 1977. Agricultural Statistics. U.S. Government Printing Office, Washington, D.C.

WILLIAMS, D.W. and CHANCELLOR, W.J. 1975. Irrigated agricultural production response to constraints in energy-related inputs. Trans. *ASAE 18* (3) 459–466.

8

Evaluation of Specific Practices

In this chapter we will examine the energetic aspects of a number of agricultural practices. Some of these practices have been proposed; others adopted. This group of practices is intended to illustrate by example how energetic considerations may influence the choice of agricultural practices.

LIMITED TILLAGE

Forms of limited tillage include no-tillage, in which less than 25% of the soil surface area is worked to plant a crop, and minimum tillage, in which the entire area may be worked but soil disturbance is limited. Limited tillage is characterized by less mechanical power for tillage and therefore lower energy requirements for tillage. Limited tillage also has numerous other characteristics, some of which affect its energetics.

Robertson and Prine (1976) and Triplett and Van Doren (1977) listed among the characteristics of limited tillage numerous advantages which may exist:

(1) Less fuel is required due to fewer and less energy-intensive field operations.
(2) Higher yields often result, particularly in dryland farming and on well-drained land.
(3) Less time and labor are required.
(4) Land use may be intensified. Double cropping potential has been extended several hundred miles northward.
(5) It is possible to farm lower quality land. Row crops may be grown on slopes to 20%.

(6) Considerably less erosion occurs.
(7) Moisture is conserved.
(8) Soil structure may be improved.
(9) There is lower investment for machinery.

There may also be disadvantages (Robertson and Prine 1976; Triplett and Van Doren 1977):

(1) Weed control depends on effective herbicides and effective management.
(2) Insects, rodents and other pests may be more serious problems.
(3) Limited tillage may not be usable with fine-textured soils.
(4) Greater amount of seed is required.
(5) Herbicides are costly.

It is evident that many farmers are finding the advantages outweigh the disadvantages. Now 2.5% of the United States cropland is farmed using no-till practices and the area farmed with minimum-tillage practices amounts to 22.5% of the United States cropland. No-till area is increasing about 6% annually and minimum-tillage about 12% annually. States with the largest areas in no-till are Indiana, Kentucky, Missouri and Pennsylvania. States with the largest areas in minimum-tillage are Kansas, Nebraska, Illinois and Iowa.

An examination of the effects of limited tillage on energy consumption centers first on the reduced fuel energy for less tillage. Table 8.1 gives energy savings found by several investigators. These savings average 1170 $MJ \cdot ha^{-1}$ and are equivalent to about 4% of the total energy requirement for growing corn. Another effect on total energy consumption is an increase due to the energy required for the herbicides; this increase is likely in the order of 200–600 $MJ \cdot ha^{-1}$. A third effect on energy consumption may occur due to lower labor requirements. Elimination of two field operations might reduce labor inputs by 1 hour per hectare or labor energy requirements by about 75 $MJ \cdot ha^{-1}$. Lower energy requirements for less machinery will be in the order of 100–200 $MJ \cdot ha^{-1}$. Overall net energy reductions for limited tillage as compared to conventional tillage may be in the order of 1000 $MJ \cdot ha^{-1}$.

Total energy requirements for United States corn production are approximately 30,000 $MJ \cdot ha^{-1}$; the net reduction due to limited tillage is about 3%. Therefore, the effect of limited tillage on energy productivity is determined primarily by yield rather than by changes in energy requirements for inputs. A minimal decrease (less than about 3%), no change, or an increase in yield will ordinarily result in increased energy productivity with limited tillage. However, limited tillage must result in increased yield if a significant increase in productivity is to be realized.

CROP DRYING

The Federal Energy Administration (FEA 1976) estimated direct energy consumption for 1974 United States crop drying as 1.0×10^{11} MJ. Much of the United States corn crop and several other crops including tobacco, soybeans, rice and peanuts are dried with energy supplied by fossil fuels, typically in the form of natural gas or LP. In 1974, crop drying was the sixth largest energy-consuming process in United States production agriculture. However, the percentage of crops being dried with heated air is on the increase. For example, 10% of the corn crop was dried in 1945, 30% in 1970, and presently over half of the crop is dried with heated air. Common practice in the past was to dry ear corn using natural air drying. Fuel consumption for tillage, planting, cultivation and harvesting of field corn typically requires about 62 $L \cdot ha^{-1}$; drying 2.5 t of corn from 25 to 15% moisture requires 46 L of LP gas. Thus, it is seen that crop drying is an area where alternative practices must be given strong consideration. Let's examine some of these alternatives.

Delaying Harvest to Provide for Additional Drying in the Field.— This alternative is difficult to evaluate since it is affected by many variables such as relative maturity of the hybrid, planting date, weather conditions, location and harvest date. For example, Morey *et al.* (1976) point out that early planting is much more effective in decreasing harvest

TABLE 8.1. REPORTED ENERGY SAVINGS WITH NO-TILL COMPARED TO CONVENTIONAL TILLAGE

Energy Savings* ($MJ \cdot ha^{-1}$)	Source
900	Heichel (1976)
1100	Alder *et al.* (1976)
1120–2430	Worsham (1977)
1240	Wittmuss *et al.* (1975)
560–1470	Robertson and Mokma (1978)
1550	Robertson and Prine (1976)
190–1920	Allen and Fryrear (1977)
740	Clark and Johnson (1975)
1630	Green *et al.* (1977)

* In most cases, the energy savings listed are due to reduced fuel consumption only and do not consider changes due to indirect energy inputs for equipment or increases for herbicides.

moisture content than delaying harvest in the fall. In addition, early planted corn shows a definite yield increase. However, other practices that result in lower harvest moisture contents, such as the use of shorter season varieties, and delaying harvest may result in significant reductions in yield. Harvest delay may be very costly in terms of additional field loss during harvest or the increased chance of crop damage by pests or bad weather. The effect of delaying harvest on the following year's crop or on multiple cropping must also be considered. The drying energy saved by delaying harvest is equivalent to less than 0.3 $t \cdot ha^{-1}$ of corn while the reduction in yield could very easily be 0.6–1.0 $t \cdot ha^{-1}$ or possibly the entire crop.

The above discussion is not intended to discourage the use of field drying, but to encourage careful planning of the planting and harvesting schedule.

Feeding High Moisture Grain.—Drying grain is not necessary if it is to be fed to livestock, particularly beef cattle. But, of course, wet grain will spoil if held for very long, depending upon its moisture content, temperature and type of storage.

High-moisture corn can be placed in an airtight storage structure and preserved through fermentation. Tight walls, coarse grinding, a plastic cover and a surface preservative such as sodium metabisulfide are necessary to exclude all air.

Propionic acid allows preservation of high-moisture grain without fermentation. Acid is sprayed on grain as it enters the bin. Bins need not be airtight, so existing storage structures can be used.

The cost of these methods of storing high moisture grain must be weighed against the cost of drying. However, it is usually possible to limit the use of heated-air drying by placing a portion of the grain aside that will be fed. It may be necessary to partially dry the high moisture grain before holding.

Use of Slower Drying Rates.—Drying at lower temperatures and slower drying rates is a practice that will save energy and dollars under some conditions. In temperate locations where grain is harvested and stored during cool weather conditions and if grain is being stored for several months anyway, low temperature (3°– 6°C above ambient) or natural air drying can be used to dry the grain over a period of several months. The cost of running the fans during this period must be considered, but periodic circulation of air through the grain is necessary anyway. Unfortunately, the use of slow drying rates in humid climates such as found in the southeastern United States cannot be used for most crops since the harvest occurs during a hot and humid portion of the year.

Under conditions where low temperature drying is not feasible, the use of high-temperature drying followed by low-temperature drying, natural air drying, or aerated stored should be considered. There is a relationship between moisture content, temperature and the length of time grain can be held without spoilage. For example, corn harvested in Florida at 25% moisture content might have a safe storage-time of one week while 20% corn could be held for one month. Thus, high temperature (fast) drying could be used to reduce the moisture content from 25 to 20%, low temperature drying (slow) could be used to further reduce the moisture content and natural air drying (no heat added) could be used to remove the last one or two percentage points down to a safe storage condition of 13–15% moisture content.

Use of Solar Drying.—Solar energy for drying grains and other crops has received considerable attention over the last few years. Most of this work has been partially funded by the U.S. Department of Energy.

Solar drying is typically accomplished by heating air with solar energy and circulating it through the product while in a storage bin. This type of drying is classified under low temperature (slow) drying as discussed above. When used in this manner, it appears to be a feasible method of drying, but the economics are still questionable. Some applications where the solar collector can be integrated into an adjacent building appear to be economically feasible at present energy costs.

The use of solar energy for high temperature (fast) drying, appears to be less promising due to the increased cost of high temperature solar collectors and the large areas required for these collectors. A recent study in Florida (Chau and Baird 1978) indicates that at least 375 m^2 of solar collectors would be required to dry 0.25 kt of corn (typical yield from 60 ha) from 25 to 15% moisture content over a one-month period.

Modifications in the Design and Operation of Existing High-temperature Dryers.—Until the energy shortages of the early 1970s, grain and crop drier designers were more concerned with drier throughput than in drier efficiency as expressed in $MJ \cdot t^{-1}$ of water removed. Therefore, only a limited number of studies on heat utilization in driers have been published. This fact alone seems to indicate that much progress can be made in more efficient use of energy in crop drying. Such parameters as drying temperature, airflow rate, drier configurations and management practices should be studied. However, changes in operating conditions for existing equipment should not be made without proper testing; i.e., Morrison (1978) reported that decreasing the drying temperature on batch or continuous-flow dryers resulted in lower (not higher) drying efficiencies.

GREENHOUSES VERSUS TRANSPORTATION

Another choice may depend in part upon whether certain winter vegetables can be produced with lower total energy consumption in greenhouses near the northern United States and Canadian markets or in production areas such as Florida, Texas, California and Mexico and transported to market. A similar choice faces Europeans. Basically the choice is determined by whether the costs of greenhouse heating or of increased transporation are greater, although other considerations such as reduced light levels at northern latitudes, different cultural practices and differences in the resulting products are pertinent.

Greenhouse heating requirements were estimated by CAST (1975) at 2380 $MJ \cdot m^{-2} \cdot yr^{-1}$ in the northern part of the United States. Leach (1976) gave two figures for the United Kingdom greenhouses; with winter lettuce 401 $MJ \cdot m^{-2} \cdot yr^{-1}$ and 536 $MJ \cdot m^{-2} \cdot yr^{-1}$. Sheard (1975) reported the total energy use for greenhouse heating in Great Britain at 1590 $MJ \cdot m^{-2} \cdot yr^{-1}$. Plantier (1977) reported French greenhouse heating requirements ranging from 190 to 1900 $MJ \cdot m^{-2} \cdot yr^{-1}$. It was reported in *Grower Talks* (Anon. 1977), a greenhouse grower publication, that heating costs from 27 growers across the United States ranged from approximately \$3–10 $\cdot m^{-2}$; with #2 oil at \$0.40 per gallon, this is equivalent to a range of 1190 to 3580 $MJ \cdot m^{-2} \cdot yr^{-1}$.

Typical production rates for tomatoes in greenhouses are 180,000–200,000 $kg \cdot ha^{-1}$ (field production rates are in the order of 28,000 $kg \cdot ha^{-1}$ for ground-grown tomatoes and up to 55,000 $kg \cdot ha^{-1}$ and above for staked tomatoes). Assuming greenhouse heating energy consumption of 2000 $MJ \cdot m^{-2}$ and production of 200,000 $kg \cdot ha^{-1}$, yields an energy requirement for heating of 100 $MJ \cdot kg^{-1}$ of tomato produced.

Using Penner's (1975) energy intensiveness of 1.65 $KJ \cdot kg^{-1} \cdot km^{-1}$ for owner-operator trucks, a full load of tomatoes would require 3.6 MJ $\cdot kg^{-1}$ to transport from Dade County, Florida to New York City or 8.3 $MJ \cdot kg^{-1}$ to transport from Sinaloa State, Mexico to New York City. It seems that from comparing these data, 10 to 25 times as much energy is required for heating greenhouses as for trucking for a given quantity of tomatoes.

Further support of the energy infeasibility of growing winter vegetables in greenhouses rather than transporting them from winter production areas was given by CAST (1977). They reported on comparisons of energy requirements for growing cucumbers and tomatoes in greenhouses versus field-grown; greenhouse cukes required 28.8 $MJ \cdot kg^{-1}$ whereas field-grown required 2.3 $MJ \cdot kg^{-1}$. Greenhouse tomatoes required 50.3 $MJ \cdot kg^{-1}$ whereas field-grown required 0.9 $MJ \cdot kg^{-1}$. With the addition of estimated transportation energy requirements to the energy requirements for field-grown production, the field-grown products yet require much less energy.

CAST (1977) indicated current research may lead to 50–70% reduction in the nighttime greenhouse heat losses. Even with reductions of this magnitude it is doubtful that energy requirements for greenhouse-grown winter vegetables can be made competitive with those of field-grown vegetables transported by truck from winter vegetable production areas.

Total costs of growing tomatoes in California and Ohio greenhouses in 1972 were reported by Cravens (1975) at $\$0.580 \cdot kg^{-1}$ and $\$0.708 \cdot kg^{-1}$. Brooke (1974) reported Florida costs were about $\$0.33 \cdot kg^{-1}$. Since then, fuel costs have increased dramatically, forcing greenhouse production costs up further, more than Florida production costs have risen. Love (1975) pointed out that greenhouse tomato growers need a $\$0.33\text{–}0.37 \cdot kg^{-1}$ premium to remain competitive with field producers. Among his component cost differences were a $\$0.22 \cdot kg^{-1}$ higher fuel cost for greenhouse and a $\$0.044 \cdot kg^{-1}$ higher transportation cost for field production. Since greenhouse production requires so much more energy than field production, as energy costs increase greenhouses will become less competitive (Plantier 1977). McElroy (1975) said in early 1975 that if fuel prices were to double again, the cost of field-grown tomatoes would increase $\$0.02 \cdot kg^{-1}$ and greenhouse tomatoes would increase $\$0.18 \cdot kg^{-1}$.

MECHANIZATION VERSUS MANUAL LABOR

In addition to yield intensive mechanization, the other result of mechanization is to reduce labor intensity. In particular, it would be useful to know the energetic results of decisions concerning the substitution of mechanization for labor.

First, let us acknowledge that one result of agricultural mechanization has been to reduce human drudgery. In general, mechanization has served to eliminate many back-breaking jobs and to replace greater numbers of minimally skilled laborers with lesser numbers of skilled laborers working under better conditions, receiving higher wages and having a higher standard of living. Humans have become less valued for their muscular output and more valued for their decision-making abilities. The quality of working conditions has been greatly improved by mechanization.

The development of an energetic basis for decisions concerning subsitution of mechanization and labor might be based upon the energy requirements of each. The energy sequestered in United States agricultural labor was estimated in Chapter 6 to be 594 $MJ \cdot d^{-1}$ or approximately 75 $MJ \cdot h^{-1}$. If the laborer is assigned to manual labor, performing strictly physical work which can be readily duplicated by a machine, he/she can be replaced by an energy input of approximately 1 $MJ \cdot h^{-1}$. Man is not competitive on an energetics basis with machinery which can replace him/her if he/she only provides manual labor.

If, instead, the worker is involved in highly skilled work or work of a decision-making or managerial character, it is difficult to estimate the energy requirement for his/her replacement. The current level of United States agricultural investment per worker of approximately $100,000 indicates the current level of substitution between capital (including land) and labor. That capital investment could generate a cash flow of about $8000 $\cdot yr^{-1}$, equivalent to an energy flow of 400,000 $MJ \cdot yr^{-1}$ or 220 $MJ \cdot h^{-1}$ of worker time. Another perhaps more realistic approach is to examine the additional capital required to supply only the mechanization or automation for worker replacement; this investment is likely no more than 3–5 times the worker's annual wages or salary. A worker receiving wages of $12,000 $\cdot yr^{-1}$ might be replaced with a capital investment of $45,000 representing a cash flow of $3600 $\cdot yr^{-1}$ or an energy flow of 100 $MJ \cdot h^{-1}$ of worker time. This capital energy requirement approximates the energy requirements for the worker, approximately 75 $MJ \cdot h^{-1}$, and indicates the generally competitive nature of such substitutions.

CROPS VERSUS LIVESTOCK

All foods are derived directly or indirectly from photosynthetic processes. The energy of sunlight is converted by photosynthesis in plants to carbohydrates which are consumed directly by man or indirectly via livestock. Gates (1971) pointed out that typical net productivities of plant communities are in the order of 1% or less of the incident solar radiation. Also, animals typically convert to body tissue about 10% of the plant energy they consume, reducing to a lower level (0.1%) the overall system efficiency (Gates 1971). Therein lies the basis for criticism of the heavy utilization of livestock in industrialized agriculture. Critics say humans should consume the feeds fed to livestock, thereby increasing system efficiency by an order of magnitude. Hannon *et al.* (1976)

determined that a soybean meat-substitute is ⅙ as energy intensive as beef. Several additional facts bear on the situation:

(1) Little of the feeds fed to livestock (forages, corn, sorghum, barley, oats) would be eaten by humans. However, on much of the cropland devoted to feed grains, food grains or vegetables, etc., could instead be grown.
(2) The quality of animal foods is high in comparison to plant products. Scrimshaw and Young (1976) showed that the human utilization of animal source proteins is high in comparison to most single plant protein sources (carefully selected combinations of plant products do, however, result in high utilization).
(3) Ruminant livestock on rangeland utilize plant production which would likely not be otherwise utilized for human food, because the land is not suitable for cultivation. About 75% of the dietary energy of cattle is roughages.
(4) Livestock products are desired by humans. The prices paid are both evidence and incentive for producers.

An important factor in the consideration of crops versus livestock is the efficiency of various livestock products. Janick *et al.* (1976) showed the efficiency of several forms of livestock products in the conversion of dietary energy in feed for livestock to products edible by humans. Eggs were highest at 18% followed by milk (17%), pork (14%), broilers (11%), beef (3%) and lamb (2%). Heichel's (1976) figures were in essential agreement, with hogs at 12%, broilers at 11%, and cattle and calves at 5%. CAST (1977) listed milk at 10–20%, broilers at 12–16%, eggs at 10–14%, pork at 5–9% and beef at 2–4%.

Pierotti *et al.* (1977) analyzed the energy sequestered in livestock products, including the energy required to produce and process the feeds consumed by the livestock. They did not include energy used directly in livestock production as did the FEA, who did not include the energy sequestered in the feed. Both sets of results, expressed in terms of energy productivity ($kg \cdot MJ^{-1}$), are therefore biased to the high side, but Pierotti and co-authors (Table 8.2) are likely the more nearly correct. CAST (1977) listed dietary energy, as a portion of total energy inputs, at 85–90% for milk and beef and 70–75% for pork, broilers and eggs.

Roller *et al.* (1975) performed a thorough process analysis on five livestock production operations, including the energy for producing feed, direct energy and the energy sequestered in buildings, equipment, labor and maintaining breeding stock. Results are in Table 8.3, expressed in kilograms of edible portion per megajoule of total energy inputs.

TABLE 8.2. ENERGY PRODUCTIVITY OF LIVESTOCK PRODUCTS

Product	Energy Productivity ($kg \cdot MJ^{-1}$)
Cattle on feed	0.011
Hogs	0.020
Turkeys	0.020
Sheep and lambs	0.021
Other beef cattle	0.022
Eggs	0.030
Chickens	0.034
Milk	0.137

Source: Pierotti *et al.* (1977).

TABLE 8.3. ENERGY PRODUCTIVITY OF LIVESTOCK PRODUCTION

Operation	Energy Productivity ($kg \cdot MJ^{-1}$)
Cow-calf thru feeder	0.023
Cow-calf feed lot	0.027
Brood sow to market hog	0.027
Ewe to market lamb	0.018
Broilers	0.025

Source: Roller *et al.* (1975).

Ward *et al.* (1977) explained the function of the United States beef production systems, showing how they developed and why they operate as they do. The feedlot system developed as a way to market grain. Beef production serves as a way to dispose of grain surpluses. Less than 10% of the total feed consumption by all beef cattle consists of feed and food grains. They estimated a range of total energy inputs for beef production from which we calculated energy productivities of 0.0093 to 0.0212 $kg \cdot MJ^{-1}$.

Williams *et al.* (1975) (Table 8.4) investigated the consumption of direct energy on three livestock operations in New York: a 40 cow dairy, a 100 cow dairy and a 1000 head beef feedlot. Included was energy consumed in the home(s) on these farms.

One thing can be predicted with certainty; the future does hold changes. With respect to livestock, we will probably see a shift away from grain-fed beef and to grass-fed beef as energy becomes more costly and less available. It is not expected that either ruminant or nonruminant livestock will exit from our agricultural systems. Ruminants have their place converting forages, rangeland vegetation and agricultural wastes, and nonruminants have the advantage of high efficiency conversion of

TABLE 8.4. ANNUAL DIRECT ENERGY CONSUMPTION ON THREE LIVESTOCK FARMS

	MJ/head		
	40 cow dairy	100 cow dairy	1000 head feed lot
Gasoline	4,790	4,880	990
Diesel	2,730	3,140	1,100
Electric power	3,280	3,190	210
Heating oil	6,280	2,510	510
Total	17,080	12,720	2,810

Source: Williams *et al.* (1975).

feeds. On a worldwide basis, crops will likely retain their predominance over livestock as a source of food for humans. Thirteen crops are produced in larger quantities than the most predominant livestock, pork, followed closely by beef (Harlan 1976). These crops in order are: wheat, rice, maize, potatoes, barley, sweet potatoes, cassava, grapes, soybeans, oats, sorghum, sugarcane and millet.

FORMS OF FOOD PRESERVATION

One alternative is in the form or degree of processing and packaging chosen. To illustrate, we will consider the energy sequestered in three commonly available forms of many foods: fresh, canned and frozen.

Differences between forms are due to the variations in the energy sequestered in packaging materials (cans are particularly energy intensive), processing energy for freezing or canning, and energy for refrigeration. The length of time stored and the storage temperature for fresh produce affects these continuing energy investments.

Herendeen and Sebald (1974) used input-output analysis to compare the effects of choice among fresh, frozen, or canned fruits and vegetables on dollar costs, energy sequestered and labor used. They found for products available in all three forms, energy consumed was about the same for fresh and frozen and somewhat less for canned. Further, fresh was most costly with frozen and canned about 80% as costly as fresh, and fresh used the most labor with frozen and canned using about 70% as much. There were, however, numerous exceptions to the averages for costs (four of 10 vegetables were cheapest fresh: greenbeans, broccoli, brussel sprouts and carrots; and four of the 10 fruits were cheapest fresh: drained peaches, drained grapefruit, drained pineapple and strawberries) and there likely were also exceptions in labor use and energy consumption.

TABLE 8.5. DIRECT ENERGY FOR PRODUCTION THROUGH HOME PREPARATION

Potatoes	Energy per unit of raw product ($MJ \cdot kg^{-1}$)	*Energy productivity ($kg \cdot MJ^{-1}$)
Fresh	7.05	0.099
Frozen	11.11	0.072

Source: Whittlesey and Lee (1976).
*Assumes 30% wastage for fresh and 20% wastage for frozen.

In an analysis on the direct energy used in production through home preparation of several foods, Whittlesey and Lee (1976) gave values pertinent to the question (Table 8.5).

Olabode *et al.* (1977) used process analysis to determine the total energy costs to deliver a 0.113 kg serving of mashed poatoes in the home using 10 intermediate forms, including fresh, frozen and canned. Their analysis was very detailed, including direct and indirect energy consumption. Results are given in Table 8.6.

Fresh produce was also found to use less energy than canned or frozen according to Pierotti *et al.* (1977). They presented figures, expressed here as energy productivity, listing several fresh fruits and vegetables ranging from 0.154 to 0.082 $kg \cdot MJ^{-1}$, and averages for canned fruits and vegetables at 0.072 $kg \cdot MJ^{-1}$, frozen fruits and vegetables at 0.050 $kg \cdot MJ^{-1}$, and dehydrated fruits and vegetables at 0.040 $kg \cdot MJ^{-1}$.

Londahl (1976) compared total energy from field to the consumer for canned versus frozen peas and found that canned was more energy intensive. However, the time the frozen product is stored before consumption would have a large effect on the outcome of the analysis.

TABLE 8.6. ENERGY SEQUESTERED IN 0.113 KG SERVINGS OF MASHED POTATOES

Processing or marketing mode	Energy productivity ($kg \cdot MJ^{-1}$)
Fresh	0.0406
Flake dried	0.0371
Granule dried	0.0367
Spray dried	0.0357
Retort pouched	0.0290
Microwave dried	0.0283
Canned	0.0235
Refrigerated	0.0223
Freeze dried	0.0164
Frozen	0.0140

Source: Olabode *et al.* (1977). Reprinted from *Journal of Food Sciences*. Copyright © by Inst. of Food Technol.

To summarize, there are probably differences in the energy productivities of foods based on the marketing form, processing and packaging. The order of energy productivities, when more than one form exists, probably is product dependent. High energy productivities will likely be favored as energy costs increase and availability declines.

Other forms of food preservation that should be analyzed with respect to energy use are drying, fermentation, pickling, concentration to high solids content (jelly, jam, condensed milk), chemical additives and radiation sterilization.

COLD PROTECTION

Energy is used in protecting crops, primarily fruits and ornamentals, from damage due to cold. Heaters are used to raise air temperature by convection and heat crops by radiation. Wind machines are used to replace cold air near ground level with warmer air from above, when a temperature inversion exists. Irrigation is used to supply sufficient water so that some remains liquid rather than freezing and the mixture remains at 0°C to prevent freezing of the crop.

Heaters consume the most energy in providing cold protection. When heaters are used, fossil fuels consumed in a single night can easily exceed all other crop production energy inputs. Oil-fired heaters are commonly spaced 10 m apart in citrus (100 heaters $\cdot$ ha^{-1}), and each usually consumes 4–8 L $\cdot$ h^{-1} or 70–145 MJ $\cdot$ h^{-1}. With some heaters adjusted to provide sufficient heat to counteract air temperatures as low as −5°C or −6°C, the fuel consumption rate per unit area can reach 30,000 MJ $\cdot$ h^{-1} $\cdot$ ha^{-1}.

Due to rising energy costs the expense of cold protection with heaters is becoming prohibitive. Growers are choosing to not fire at approximately −3°C to protect fruit but instead at −5°C to protect the trees, sacrificing a crop of fruit.

What are the energetic aspects of choices concerning cold protection? The loss of a single crop is the loss of the energy invested in producing that crop, which is a year's production energy inputs plus, for a grove or orchard, a portion of the energy inputs for its establishment. On an energetic basis, it would not be reasonable to consume for cold protection more than the energy sequestered in a crop in order to protect the crop; it would require less energy to grow another crop in another year. For instance, if the annual energy requirements, including amortization of grove establishment, for citrus are 36,000 MJ $\cdot$ ha^{-1}, only two hours of cold protection requiring 18,000 MJ $\cdot$ ha^{-1} $\cdot$ h^{-1} could be justified on an energetic basis before the crop would be written off. It should be noted that, even with current high energy costs, economics would justify several additional hours of cold protection rather than sacrificing the crop.

GREEN REVOLUTION

World populations of 4 billion (present) and 7 billion projected for the year 2000 and 16 billion projected for the year 2135 give some idea of the challenges that face mankind to feed a rapidly growing population.

Most people of the world desire to eat and live as we do in the United States. Hence, one consideration is to feed a population of 4 billion with a United States high protein-calorie diet produced with the use of United States agricultural technology. Presently, about 160 million ha are planted to crops in the United States. With about 210 million people in the United States, this gives about ¾ ha of crop per capita. Since about 20% of our crop yield is exported, the estimated arable land per person is about 0.6 ha.

World arable land area is about 1.5 billion ha. With 4 billion humans in the world today, that gives only 0.38 ha per person worldwide. In the United States 0.6 ha of land plus a high-energy agricultural technology are necessary to produce the high protein diet that we consume. Hence, there is not enough arable land in the world today (even assuming that the energy resources and other technology were also available) to feed the current world population of 4 billion a diet similar to that consumed in the United States. Figure 8.1 gives a visual representation of the world's arable land by sizing each major land mass in proportion to its potentially arable land. The silhouette map within each outline shows the portion of that arable land that was being cultivated in the mid 1960s.

In the above analysis, fossil energy was assumed to be unlimited. Unfortunately, fossil fuel is limited for food production. If the world were fed by a food production system similar to the United States', approximately 3.1×10^{12} L of fuel would be required annually. The known world reserves of petroleum has been estimated to be 57×10^{12} L. If we assume 75% of the raw petroleum can be converted into fuel, this would amount to approximately 43×10^{12} L of usable fuel. If petroleum were the only source of energy for food production and if we used petroleum only for producing food, the world petroleum reserves would last about 14 years.

Thus, it is doubtful whether the so-called "Green Revolution"—the scientific and technological advances resulting in greater production per unit of land—can solve the world problem of food shortage. To further illustrate this point, only in two instances, namely wheat in India and

From Hopper (1976) with permission of Scientific American

FIG. 8.1. COMPARISON OF WORLD'S POTENTIALLY ARABLE LAND WITH THAT UTILIZED IN MID 1960s

The outline maps show the world's major landmasses sized in proportion to the area of their potentially arable land. The silhouette map within each outline shows how much of that potentially arable land was being cultivated as of the mid 1960s. The numbers give the cultivated area as a percentage of the potentially arable area.

Mexico, have yields managed to outrun population gains; this feat was accomplished primarily through irrigation (Borgstrom 1973).

On the other hand, Hopper (1976) takes a more optimistic view and indicates that the Green Revolution demonstrates that great accomplishments can be made with developing countries in only a few years. He refers to the 200 and 300% increases in yields with the use of improved varieties of wheat and rice in South Asia during the past few years. He pointed out that the Asian peasant proved to be as innovative as any farmer in the world and that the idea of the traditional farmer being slow to change and resistant to progress died a much-deserved death.

Hopper also suggests that the world's food problem does not arise from any physical limitation on potential output or any danger of unduly stressing the "environment"; but rather on the social and political structures of nations and their economic relations.

Two-thirds of the world population live in countries with national average diets that are nutritionally inadequate. A country is classified as diet-deficient if the average annual per capita consumption of food results in a deficiency of calories, proteins, or fat below minimum levels recommended by nutritionists. The diet deficient areas include all of Asia, except Japan, Israel and the Asian part of the U.S.S.R., and all but the southern tip of Africa, part of South America, and almost all of Central America and the Caribbean (Abel and Rojko 1967). Note that the problem is mostly in the tropics, subtropics and contiguous areas. According to Thurston (1969), failure to recognize the world food shortage as a tropical problem often leads people to believe that our temperate zone agricultural technology can be transplanted directly. Even if energy shortages were not a problem much research would be necessary to develop a productive, efficient food producing system under tropical conditions.

Rice, wheat, maize, sorghum, the millets, rye and barley are principal foods in cereal-producing areas, but elsewhere cassava, the sweet potato, potatoes, coconuts and bananas are basic foods. Although over 3000 plant species have been used for food and over 300 are widely grown, only about 12 furnish nearly 90% of the world's food. Many more of those crop plants are grown in the tropics than in the temperate zones,and we in the temperate zone often do not know much about these plants. The fact is that very little research has been done on food producing plants in the tropics. Starchy crops such as cassava, sweet potatoes, yams and plantains hardly enter into world commerce when compared to cereals. One problem is storage of these products. Also, nutritionists generally dislike these crops because they are primarily carbohydrate and have a low protein content, yet varieties exist with an appreciable protein content and much might be done in a breeding program to increase it (Thurston 1969).

BIBLIOGRAPHY

ABEL, M.F. and ROJKO, A.S. 1967. World food situation. Prospects for world grain production, consumption, and trade. Foreign Agric. Econ. Rept. No. *35.* U.S. Dep. Agric., Washington, D.C.

ADLER, E.F., KLINGMAN, G.C. and WRIGHT, W.L. 1976. Herbicides in the energy equation. Weed Sci. *24,* 99–106.

ALLEN, R.R. and FRYREAR, D.W. 1977. Limited tillage saves soil, water, and energy. ASAE Paper No. *77-2029.* Ppr. presented at 1977 Ann. Meet. Amer. Soc. Agric. Eng., St. Joseph, Michigan.

ANON. 1977. 27 Growers' fuel costs and why. Grower Talks *40* (12) 30–37.

ANON. 1978. Tillage practices: What are farmers doing? Implement & Tractor *93* (9) 44–46.

BAKKER-ARKEMA, F.W., DEBOER, S.F. and LEREW, L.E. 1973. Energy conservation in grain driers: I. Performance evaluation. ASAE Paper No. *73-324.* Ppr. presented at 1973 Ann. Meet. Amer. Soc. Agric. Eng., St. Joseph, Michigan.

BORGSTROM, G. 1973. The Food and People Dilemma. Duxbury Press, Belmont, California.

BROOKE, D.L. 1974. Costs and Returns from Vegetable Crops in Florida, Season 1972–73 With Comparisons. Econ. Rept. *59,* Food and Res. Econ. Dep., Univ. of Fla., Gainesville.

CHAU, K.V. and BAIRD, C.D. 1978. Solar grain drying under hot and humid conditions. Proc. 1978 Solar Grain Drying Conf. Purdue University, West Lafayette,Indiana.

CLARK, S.J. and JOHNSON, W.H. 1975. Energy-cost budgets for grain sorghum tillage systems. Trans. ASAE *18* (6) 1057–1060.

COUNC. AGRIC. SCI. TECHNOL. (CAST). 1975. Potential for energy conservation in agricultural production. Rept. No. *40.* Counc. for Agric Sci. and Technol., Dep. of Agronomy, Iowa State Univ., Ames.

CAST. 1977. Energy use in agriculture: Now and for the future. Rept. No. *68,* Counc. for Agric. Sci. and Technol., Dep. of Agronomy, Iowa State Univ., Ames.

CRAVENS, M.E. 1975. Competition and industry development. *In* Tennessee Valley Greenhouse Vegetable Workshop, March 18–20. U.S. Dep. Agric. and Tenn. Valley Authority. Chattanooga, Tenn.

COMM. AGRIC. PRODUCTION EFFICIENCY. 1975. Agricultural Production Efficiency. Natl. Acad. Sci., Washington, D.C.

FED. ENERGY ADMIN. (FEA). 1976. Energy and U.S. Agriculture/Data Base. Fed. Energy Admin.—U.S. Dep. Agric., Washington, D.C.

FEA. 1977. A guide to Energy Savings for the Field Crops Producer. Fed. Energy Admin.—U.S. Dep. Agric., Washington, D.C.

GATES, D.M. 1971. The flow of energy in the biosphere. Sci. Amer. *224* (3) 88–92, 94, 96–100.

GREEN, M.B., HARTLEY, G.S. and WEST, T.F. 1977. Chemicals for Crop Protection and Pest Control. Pergamon Press, New York.

HALL, F.W., MADDEX, R.L. and BAKKER-ARKEMA. F. 1975. Energy utilization on selected Michigan farms. ASAE Ppr. No. *75–3017*. Ppr. presented at Ann. Meet. Amer. Soc. Agric. Eng., St. Joseph, Michigan.

HANNON, B.M., HARRINGTON, C., HOWELL, R.W. and KIRKPATRICK, K. 1976. The dollar, energy and employment costs of protein consumption. CAC Doc. No. *182*, Cntr. for Adv. Computation, Univ. of Ill., Urbana.

HARLAN, J.R. 1976. The plants and animals that nourish man. Sci. Amer. *235* (3) 89–97.

HEICHEL, G.H. 1976. Agricultural production and energy resources. Amer. Sci. *64* (1) 64–72.

HERENDEEN, R. and SEBALD, A. 1974. The dollar, energy, and employment impacts of certain consumer options, Vol. 1. CAC Doc. No. *97*, Cntr. for Adv. Computation, Univ. of Ill., Urbana-Champaign.

HOPPER, W.D. 1976. The development of agriculture in developing countries. Sci. Amer. *235* (3) 197–205.

JANICK, J., NOLLER, C.H. and RHYHERS, C.L. 1976. The cycles of plant and animal nutrition. Sci. Amer. *235* (3) 75–84, 86.

LEACH, G. 1976. Energy and Food Production. IPC Science and Technology Press, Guildford, Surrey, UK.

LONDAHL, G. 1976. Energy requirements: Freezing vs canning. Food Eng. *489* (9) 55–56.

LOVE, H.G. 1975. Economic considerations before entering the greenhouse vegetable business. *In* Tennessee Valley Greenhouse Vegetable Workshop, March 18–20, U.S. Dep. Agric. and Tenn. Valley Authority, Chattanooga, Tenn.

MCELROY, R.G., II. 1975. The competitive position of greenhouse vegetables: A study of rising fuel costs. *In* Tennessee Valley Greenhouse Vegetable Workshop, March 18–20. U.S. Dep. Agric. and Tenn. Valley Authority, Chattanooga, Tenn.

MOREY, R.V., CLOUD, G.A. and LUESCHEN, W.E. 1976. Practices for efficient utilization of energy for drying corn. Trans. ASAE *19* (1) 151–155.

MOREY, R.V., GUSTAFSON, R.J., CLOUD, H.A. and WALTER, K.L. 1978. Energy requirements for high-low temperature drying. Trans. ASAE *21* (3) 562–567.

MORRISON, D.W. 1978. Energy management in grain drying. Grain drying and energy management workshop, Dec. 4. Chicago.

OLABODE, H.A., STANDING, C.N. and CHAPMAN, P.A. 1977. Total energy to produce food servings as a function of processing and marketing modes. J. Food Sci. *42* (3) 768–774.

PEART, R.M. and LIEN, R.M. 1975. Grain dryer energy requirements. ASAE Paper No. *75–3019.* Ppr. presented at Ann. Meet. Amer. Soc. Agric. Eng., St. Joseph, Michigan.

PENNER, P.S. 1975. The Dollar, Energy and Labor Impacts of 1971 Motor Freight Transportation. CAC Tech. Memo. No. *44,* Cntr. for Adv. Computation, Univ. of Ill., Urbana-Champaign.

PIERCE, R.O. and THOMPSON, T.L. 1975. Energy utilization and efficiency of crossflow grain dryers. ASAE Paper No. *75–3020.* Ppr. presented at Ann. Meet. Amer. Soc. Agric. Eng., St. Joseph, Michigan.

PIEROTTI, A., KEELER, A.G. and FRITSCH, A.J. 1977. Energy and Food. CSPI Energy Series X, Cntr. for Sci. in the Public Interest, Washington, D.C.

PLANTIER, R. 1977. The use of energy in European agriculture. Monthly Bull. Agric. Econ. and Statist. *26* (6) 1–9.

ROBERTSON, L.K. and MOKMA, D.L. 1978. Crop residue and tillage consideration in energy conservation. Energy Fact Sheet *6,* Mich. State Univ., East Lansing.

ROBERTSON, W.K. and PRINE, G.M. 1976. Conserving energy with no-tillage. EC-37, Univ. of Fla., Gainesville.

ROLLER, W.L., KEENER, H.M. and KLINE, R.D. 1975. Energy Costs of Intensive Livestock Production. ASAE Paper No. *75–4042.* Ppr. presented at Ann. Meet. Amer. Soc. Agric. Eng., St. Joseph, Michigan.

SCRIMSHAW, N.S. and YOUNG, V.R. 1976. The requirements of human nutrition. Sci. Amer. *235* (3) 51–64.

SHEARD, G.F. 1975. Energy requirements for glasshouse heating in Britain. Span *18* (1) 27–28.

SHOVE, G.V. 1973. Low temperature drying. Proc. of Univ. of Ill. Grain Conditioning Conf. Low Temperature Drying and Chemical Preservations. Univ. of Ill., Urbana.

SINGH, P.R. 1977. Energy consumption and consideration in food sterilization. Food Technol. *31* (3) 57–60.

SMERDON, E.T. 1977. The role of energy in food production—A perspective. *In* Energy Use Management. R. Fazzolare and C.B. Smith (Editors). Pergamon Press, New York.

THURSTON, D.H. 1969. Tropical agriculture—A key to the world food crisis. Bioscience *19* (1) 29–34.

TRIPLETT, G.B., Jr. and VAN DOREN, D.M., Jr. 1977. Agriculture without tillage. Sci. Amer. *236* (1) 28–33.

UNGER, S.G. 1975. Energy utilization in the leading energy-consuming food processing industries. Food Technol. *29* (12) 33–45.

WARD, G.M., KNOX, P.L. and HOBSON, B.W. 1977. Beef production options and requirements for fossil fuel. Science *198,* 265–271.

WHITTLESEY, N.K. and LEE, C. 1976. Impacts of energy price changes of food costs. Bull. *822,* Coll. of Agric. Res. Cntr.,Washington State Univ., Pullman.

WILLIAMS, D.W. *et al.* 1975. Energy utilization on beef fed lots and dairy farms. *In* Energy, Agriculture and Waste Management. W.W. Jewell (Editor). Proc. 1975 Cornell Agric. Waste Mgt. Conf., Ann Arbor Sci. Publishing Co., Ann Arbor, Michigan.

WITTMUSS, H., OLSON, L. and LANE, D. 1975. Energy requirements for conventional versus minimum tillage. Soil & Water Conserv. *30* (2) 72–75.

WORSHAM, A.D. 1977. No-till: Worth trying. Weeds Today April–May, 16, 18–19.

9

New Sources of Energy

NONRENEWABLE SOURCES

In recent years, it has become common knowledge that our oil and gas reserves are being depleted at an alarming rate. Since our food system, and particularly production agriculture, relies heavily on these fuels, future shortages will probably have a greater affect on food production than on other sectors of the economy.

Coal is unquestionably the fossil fuel of the future, particularly for the United States. However, as S. David Freeman has said, referring to environmental problems, "there are two things wrong with coal today. We can't mine it and we can't burn it." The challenge is to find new ways of using coal to satisfy our energy needs. Even if pollution were not a problem, it would be difficult to utilize coal directly in most agricultural systems, since they normally require easily portable fuels such as liquid petroleum. Thus, the development of new uses of coal that will be compatible with food production systems is very important.

New Uses of Coal

Gasification offers one possible solution to using coal to produce a more desirable fuel. The basic chemistry of gasification is simple. Carbon from coal or naphtha—the petroleum fraction with a boiling point between 125° and 240°C is combined with water at a high temperature to form methane, the principal constituent of natural gas. The overall reaction is much more complex and involves several steps.

Naphtha gasification is considerably simpler than coal gasification and is in a much more advanced state of development. There are three processes that are in commercial use. The major problem is the high cost, as much as five times (or more) the cost of natural gas.

Gas made from coal was once widely used in the United States. It has been only about 25 years since United States utilities shifted from coal gas to natural gas. However, the processes used to produce coal gas at that time were very inefficient and polluting.

Other new methods of using coal include combined power gas and steam cycle systems. Power gas is a low-heat content gas from coal, and because of its low-heat value it cannot be economically transported long distances. Another more efficient method of using coal is with a magnetohydrodynamic (MHD) generator, which converts heat from combustion gases directly into electricity. It not only requires less fuel and produces less thermal pollution, but also offers one of the best methods of eliminating sulfur oxide and reducing nitrogen oxide emissions from coal-fired power plants.

Despite its promise, MHD has yet to be proven as a practical technology, partly because support for construction of large-scale facilities has not been made available. The MHD generator is basically an expansion engine in which hot, partially ionized gases flow down a duct lined with electrodes and surrounded by coils that produce a magnetic field across the duct. Thus, the conducting gas passing through the magnetic field generates a current that is collected at the electrodes and requires no moving parts.

Nuclear Energy

Nuclear energy, if it can be produced by relatively safe and nonpolluting methods, may be the long-term answer to the energy problem. Maybe more important, it could be the answer to the world's food problem. As has been pointed out, it is very difficult to increase yield without increasing energy input. A very small percentage, if any, of the world's arable land has reached its maximum yield; yields have been primarily limited by economic considerations and shortages of inputs. Abundance of electrical power might render production of hydrogen by electrolysis feasible, thus providing agriculture with fuel and other inputs such as fertilizer.

Fission Technology.—Man's first use of nuclear energy was for weapons. The establishment of the atomic energy commission in the late 1940s had as its goal to develop peaceful benefits from the atom, but progress has been slower than expected.

Uranium (and potentially thorium) are the raw materials of nuclear energy. The uranium fuels used in existing reactors release about 20,000 times as much heat as can be obtained from the equivalent weight of coal; and in the more efficient breeder reactors now being developed, this ratio will be as high as 1,500,000 to one. Of course, the handling of radioactive wastes and other special safety precautions present the major drawbacks.

The uranium found in nature is mostly ^{238}U and contains only about 0.7% ^{235}U, the isotope that undergoes fission releasing heat within a reactor. For present-day United States reactors, (mostly light water reactors , LWR) the fuel must be enriched by increasing the concentration of ^{235}U. This results in a very inefficient use of our uranium supply, and if used in this manner the known resources would produce less energy than our coal reserves. For this reason, LWRs were never considered as more than a stop-gap by the early promoters of nuclear power. If fission is to become a major source of energy, breeder reactors will be needed. Breeders produce more fissionable material than they consume and thus theoretically can utilize between 50 and 80% of the uranium and thorium resources.

The present system requiring enrichment of uranium not only wastes our nuclear fuels but also is very dependent upon strip-mined coal to supply power for the enrichment process.

The U.S.S.R. already has a prototype breeder reactor in operation while France and Britain are nearing completion of similar plants. Numerous designs of fast breeder reactors are under investigation.

Fusion Technology.—In the early atomic age it was realized that the reaction that produces the hydrogen bomb could be a great source of energy if it could be controlled. At one time, it was thought that research on a fusion reactor might proceed so quickly that fusion would become an alternative to the first generation of breeder fission reactors. However, progress has been slow because a whole new field of science must be mastered—plasma physics.

Magnetic containment fusion appears to be promising, with the possibility of controlled fusion within the next few years. A new approach to fusion using a laser to heat the fuel may be even more promising. Of course, even if either magnetic or laser fusion were demonstrated, it would take many years before it would be commercially available. It is clear that fusion reactors would have two advantages: virtually unlimited fuel resources and no conceivable danger of an explosive accident.

Two heavy isotopes of hydrogen, deuterium and tritium, are commonly considered as the likely fuels for fusion. Deuterium is so plentiful in seawater that it would be an extremely cheap fuel (costing only 0.003 mill[1] per kilowatt hour) and could provide the world with power for a million years or more. However, tritium would have to be bred in a fusion reactor, much like plutonium. As an ultimate resource, the energy available from fusion is surpassed only by the resource of sunlight, which of course is just fusion power from a far greater reservoir.

The temperature for fusing deuterium and tritium is so high that no material could contain the fuel without melting. Thus, magnetic fields are being investigated as possible containers with several designs being considered.

Laser fusion takes a different approach in that a small, well-focused pulse of laser light very rapidly heats a small pellet of deuterium and tritium to temperatures hot enough to ignite fusion (100 million°C) before the pellet can expand significantly. Because Newton's law of inertia largely determines the speed of expansion (about 1000 $km \cdot s^{-1}$), laser fusion is often called inertially confined fusion. A 1 mm pellet expanding at 1000 $km \cdot s^{-1}$ would double in size in 1 ns, so the laser energy must be delivered very rapidly.

RENEWABLE RESOURCES

Geothermal Energy

The earth's heat is a potentially valuable source of energy. Many scientists believe this source of heat can be used to generate substantial amounts of electricity in the near future. Three types of resources are being considered: steam, hot water and hot rock. Estimates reported by McCloskey (1978) of potential development by 1985 range from 4000 MW to 132,000 MW and by 2000 from 30,000 MW to 800,000 MW. The range in estimates indicates that from 0.5 to 15% of the country's electric power might come from geothermal sources. Presently, only 0.1% comes from that source, mostly in California. Considering California alone, it is estimated that 20–30% of its electric power could be supplied from geothermal by 2000.

Other forms of geothermal energy are being considered such as cool air from caves, old mines and underground reservoirs not filled with water. For example, constant temperature air from abandoned coal mines in Kentucky is being investigated for use in greenhouses (Buxton *et al.* 1977).

---

[1] One mill equals 1/10 cent.

Solar Energy

Solar energy is the most abundant form of energy available to man, and is so plentiful that the energy arriving on 0.5% of the land area of the United States is more than the projected total energy needs of the country in the year 2000. However, sunlight is diffuse and intermittent and will require large investments in collection and storage systems.

For example, consider the land area mentioned above, which amounts to about 6 million ha, and current costs and efficiencies of solar collection systems ($100 · m^{-2} and 50% efficiency for low-temperature applications). Thus, it would require 12 million ha of collectors at a total of over $13 trillion, much larger than our present GNP. Also, many applications such as electrical power generation require much more expensive solar collectors and have system efficiencies of 5–15%. One would not expect the cost of conventional solar collectors to become much cheaper, since any reduction in cost due to mass production is likely to be offset by increased material costs as energy and cost-of-living expenses increase.

Despite its abundance, solar energy has not been expedited except in a limited way in water heaters, furnaces and space applications. The technologies that would allow more widespread use, such as direct conversion to electricity or refrigeration, are not commercially available. Whether or not solar energy becomes generally available in the near future, there is a growing agreement that it will be important in the long run.

For space heating, the solar collector is typically a black surface that readily absorbs sunlight and is covered with a light-transmitting material such as glass or plastic to reduce heat losses from the absorbing plate. The heat absorbed by the metal plate is transferred by moving air or water and is usually stored for later use in insulated hot water tanks, rock beds or some phase change material.

Solar air conditioning and refrigeration systems will likely be developed similar to the gas refrigeration systems utilizing an absorption cycle. The heat supplied by the gas flame will be replaced by solar heat, probably in the form of hot water. Several experimental systems using this principle are being tested.

Combined cooling and heating systems, which have not yet been developed commercially, are expected to improve the economic prospects for both because of the joint use of the collector.

In most regions of the country backup systems based on conventional fuels will be needed for extended periods of bad weather. Nevertheless, it has been estimated that if systems were commercially available now, solar water and space heating would be cheaper than electric heating in nearly all of the United States and would be competitive with gas and oil heating when these fuels double in cost (ERDA 1976). It is believed that solar heating and cooling systems could ultimately supply as much as half of the 20% of the total United States energy consumption that is now used for residential and commercial buildings.

Generating electricity with heat from solar energy is a more difficult problem, and there are conflicting ideas about the best approach to the problem. Some engineers believe that small generating units located where the electrical power is to be used are ideal since solar energy is diffuse and well distributed. Others prefer large centralized facilities.

A program to develop solar energy in agricultural applications is funded by the U.S. Department of Energy and administered through the U.S. Department of Agriculture. This program includes solar crop drying, solar grain drying, solar energy use in livestock production, solar greenhouse heating and solar energy use in agricultural food processing (see Bibliography for program reports and proceedings). Many of these research projects are developing low-cost components such as collectors constructed from plastic materials.

Significant progress has been made toward making solar energy applications in agriculture economically feasible; however, very few applications can clearly be defined as such. The primary problems with respect to economic feasibility are that solar energy applications in agriculture are competing with relatively low-cost fossil fuels and many agricultural applications, such as crop drying, require energy only for a small portion of the year. Taking crop drying as an example and assuming that the system will operate two months a year with a 50% solar collection efficiency, results in an annual fuel savings of about $20\ L \cdot m^{-2}$. If the system could be used 12 months a year or if solar energy were used to replace electric heat for crop drying, the economics of solar drying would be very competitive. Some electric heat is being used for low temperature grain drying in the corn belt and offers an opportunity for cost effective solar drying. Grain drying is one of the few commercially adopted solar energy processes used in United States agricultural production.

A report prepared for the National Science Foundation (1976) on Solar Energy Applications in Agriculture gave the following observations and recommendations:

"(1) Heat requirements in terms of temperature for many of the processes studied are relatively modest.

"(2) Basic technology already exists for generating the required level of heat from solar sources.

"(3) Storage of heat is required for many of the processes.

"(4) Management methods will require modification to make effective use of solar energy.

"(5) Irrigation essentially requires shaft power which is an area of solar energy technology different from those for producing heat alone that has yet to reach practical development.

"(6) Applications of solar energy technology to meet specific requirements of each end use through research, design, development and demonstration are the keys to early creation of both solar energy components and systems. These require enlarged and more heavily funded R,D&D programs.

"(7) A variety of strategies and incentives which focus on utilization of the existing agricultural infrastructure operating in concert with other public agencies, private industry and educational institutions can expedite acceptance and use of solar energy on the farm. This is especially true of the need for high initial capital investment in solar energy components and systems."

Despite the advantages of solar energy, acceptance may be slow unless what are essentially societal and monetary problems can also be solved. The building industry is traditionally slow to adopt new techniques, and solar systems do require much higher initial investments. The higher initial investment is a definite deterent to the adoption of solar energy applications in agriculture, since farmers are reluctant to make long-term investments on equipment that is specific to a given crop or application. Tax incentives and long-term, low-interest loans would be a big boost.

Fuel from Wastes

As the United States consumes greater quantities of fossil fuels than it can produce it has also begun to produce far greater quantities of solid organic wastes than can be consumed by landfills and other conventional methods of disposal. The obvious approach seems to be to convert this waste into synthetic fuels, helping to solve both problems.

Many arguments in favor of this approach have been highly simplistic and have ignored the difficulties of marketing low-value energy resources. The proponents of conversion, furthermore, have frequently both overestimated the amounts of suitable waste materials available and underestimated the difficulty of collecting sufficient quantities at a given location to make the operation economically feasible.

The United States generates tremendous quantities of solid wastes—about 1 Gt (gigatonne) of inorganic mineral wastes and more than 2 Gt of organic wastes each year. These figures are somewhat misleading as to the potential energy source since a recent study indicates half the total weight is water and more than 80% of the total was so widely dispersed that it could not be used (Table 9.1). However, the development of successful techniques for using wastes will certainly accelerate their collection.

There are three major routes for conversion of these wastes to synthetic fuels: hydrogenation, pyrolysis and bioconversion. Hydrogenation and pyrolysis have been brought to the pilot and demonstration plant stages of development and will probably be commercialized within a few years. Bioconversion has received only minor research effort until recently and probably will not be commercialized as soon.

The hydrogenation process developed by Appell *et al.* (1970) might be termed deoxygenation since the principal reaction is abstraction of oxygen from cellulose—the primary component of organic wastes. In the process, organic waste and as much as 5% of an alkaline catalyst such as sodium carbonate are placed in a reactor with carbon monoxide and steam at an initial pressure of 100 to 250 atmospheres and heated at 240° to 380°C for as long as one hour. Under optimum conditions, as much as 99% of the carbon is converted to oil—about 250 $L \cdot t^{-1}$ of dry waste. In practice, more than 85% conversion is normally obtained, but, subtracting the heat used in the process, the actual yield is about 150 $L \cdot t^{-1}$ of dry waste. The product is a heavy paraffinic oil with a heat value of 35 $MJ \cdot kg^{-1}$. In comparison, the widely used No. 6 fuel oil has a heat value

TABLE 9.1. AMOUNTS (t) OF DRY, ASH-FREE ORGANIC SOLID WASTES PRODUCED IN THE UNITED STATES IN 1971

Source	Wastes generated	Readily collectable
Manure	181	24
Urban refuse	117	64
Logging and wood manufacturing residues	50	5
Agricultural crops and food wastes	354	21
Industrial wastes	40	5
Municipal sewage solids	11	1
Miscellaneous	45	5
Total	798	125
Net oil potential (TL)	12.2	1.9
Net methane potential (GL)	249	39

Source: Anderson (1972).

of 42 MJ · kg^{-1}. The energy value of the raw waste varies from 7 to 19 MJ · kg^{-1}.

The second major route for production of synthetic fuels is pyrolysis (destructive distillation). A major disadvantage of pyrolysis is that the process generally produces at least three different fuels—gas, oil and char—thus increasing the recovery and marketing problems. However, the construction and operating costs should be lower than for hydrogenation.

The Garrett pyrolysis process (Hammond *et al.* 1973) is part of a complete system for disposal of urban refuse. Waste is first shredded and dried, and inorganic materials are removed for recycling or disposal. The organic waste is then reshredded and heated to about 500°C in an oxygen-free atmosphere. Each tonne of waste produces about 125 L of oil, 63 kg of ferrous metals, 54 kg of glass, 73 kg of char and a varying amount of low energy gas (0.15 to 0.15 MJ · L^{-1}).

Methane can be produced by the third major technology, bioconversion through digestion by anaerobic bacteria. Development of this process is less advanced than the others, but current work suggests that about 40,000L of methane, with an energy content of about 3.7 MJ · L^{-1}, could be produced from each tonne of solid waste.

Although bioconversion is theoretically a simpler process than hydrogenation or pyrolysis, a large number of problems remain to be solved. Among them are the need for new techniques to feed solids into the digestors and inexpensive ways to collect and purify the methane, recirculate the effluents and control pollution. A major environmental problem is disposal of the organic sludge that remains after digestion, which may amount to 40% of the starting material.

This sludge could possibly be dried and burned, but that would produce air pollution. It could be converted to oil or gas with one of the other techniques, but it would seem more logical to use that technique for all the waste. Or since the sludge has a high protein content, it might prove valuable as an animal feed additive.

The economics of sludge disposal will play a major role in determining the overall feasibility of bioconversion. It appears that bioconversion is more applicable to agricultural wastes and probably will find a good use for the sludge produced.

Growing Energy Crops

Recent proposals have been made in which a major portion of a country's petroleum needs would be supplied by cultivating specific crops. The Brazilian government has a goal of supplying 20% of its total fuel consumption (30 billion L in 1977) by 1980 in the form of alcohol from sugar cane (Stout *et al.* 1978). Brazil produced about 2 billion L of fermentation alcohol in 1976, so their goal is not completely unreasonable. The Brazilian automobile industry has modified the internal combustion engine and added a heat exchanger so it can use the alcohol fuel with no loss of efficiency. These are American, German and Japanese cars which are modified on assembly in Brazil.

Can the United States use the same approach as Brazil? The United States cannot afford to use farmland for fuel as Brazil does, but kelp could possibly be grown off the California coast or the waste cellulose in forestry operations could be hydrolyzed to glucose, which then could be converted into sugar. However, the United States annual biomass production for food, lumber, paper and fiber, if exclusively used for energy, would provide only 25% of our present energy requirements (Burwell 1978). Another problem is the higher cost of fuel produced from biomass (two or three times current gasoline prices).

Other possibilities are to grow crops that produce hydrocarbons via photosynthesis. Rubber growers in Malaysia have produced 4500 kg $\cdot$ ha^{-1} on experimental plots. One tonne of hydrocarbon equals about 1220 L of oil. Guayule, a desert shrub in the western United States, produces a high molecular weight hydrocarbon. The uncultivated plant yield represents not less than 780 $L \cdot ha^{-1} \cdot yr^{-1}$. This could go as high as 4000 $L \cdot ha^{-1} \cdot yr^{-1}$. The Gopher plant or mole plant grows to a height of about 1 m and contains hydrocarbon-like material in its leaves.

Fast-growing algae in ponds or the ocean seem to be the best approach for producing more biomass for conversion. Much higher photosynthetic efficiencies (4 to 10%) can be attained with algae, compared to conventional farming (rarely above 0.5%). Single-cell algae in fresh water with a high nutrient content exhibit growth rates as high as 190 (dry) $t \cdot ha^{-1} \cdot yr^{-1}$ with 44 to 88 (dry) $t \cdot ha^{-1} \cdot yr^{-1}$ expected for a 365–day growing season. Water hyacinth in the southeastern United States can grow up to 160 (dry) $t \cdot ha^{-1} \cdot yr^{-1}$ (Greeley and Spewak 1976).

Because of the limited availability of fresh water suitable for farming, ocean farming is being considered. One consideration is the giant California kelp, which converts solar energy to biomass energy with a 2% efficiency under proper growing conditions. Using anaerobic digestion to extract methane from the kelp, a 575,000 km^2 farm would supply the United States demand for natural gas.

Another possible source of energy, although not biological photosynthesis, is based on what we know about the photosynthetic process. It is a "synthetic" system, in which we expect to sensitize the photodecomposition of water to hydrogen and oxygen (Calvin 1977). Hydrogen is a fuel that can be used in various ways.

Wind Energy

Thirty or forty years ago, wind-generated power played an active part in the electrification of many rural homesteads across the country (Clews 1974). This soon ended with the advent of the Rural Electrification program.

There is sufficient wind energy available in the United States to provide about 14% of the projected energy demand in the year 2000. However, this would require the installation of over 300,000 large and very expensive wind energy conversion systems (WECS). A more realistic estimate of the wind energy contribution in 2000 is between 0.5 and 7.0%

Adequate levels of wind can be found over half of the United States land area. However, the most attractive high energy areas (average wind speed over 24 km · hr^{-1}) would produce only 4% of the potential wind energy available in the United States. Figure 9.1 shows the favorable wind regimes in the United States. Note that very few favorable areas exist in the eastern United States.

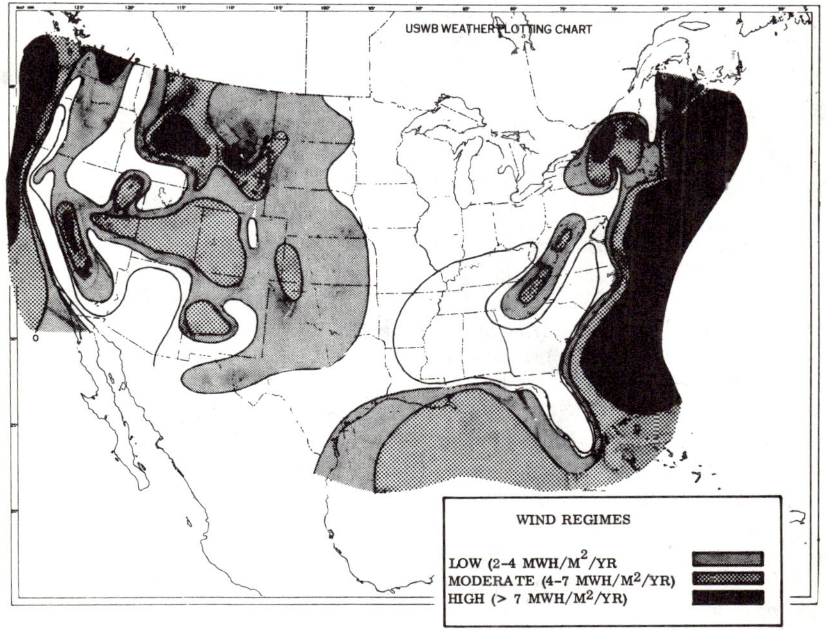

FIG. 9.1. FAVORABLE WIND REGIMES

TABLE 9.2. AVAILABLE POWER FROM THE WIND IN RELATION TO AVERAGE WIND SPEED AND PROPELLER DIAMETER IN A SYSTEM WITH 40% OVERALL EFFICIENCY

Propeller diameter, m	Output power in watts for average wind velocity, $km \cdot h^{-1}$				
	8	16	24	32	40
1.2	3	27	90	213	416
2.4	13	107	360	853	1665
3.7	30	240	809	1918	3746
4.9	53	426	1439	3410	6660
6.1	83	666	2248	5328	10407

Source: Soderholm and Andrews (1974).

Table 9.2 shows the available power from the wind in relation to average wind speed and propeller diameter. Very few locations in the eastern United States have an average wind velocity of greater than 8 $km \cdot h^{-1}$, meaning that a 6.1 m diameter unit would produce less power than that required by a 100-watt light bulb, while costing about $5000. In other words, wind power has very little potential in most of the eastern United States. Applications such as pumping water in remote areas may be more attractive; however, one should carefully consider the economics as well as the energetics. It is very likely that a windmill operating in the eastern United States would produce less energy in its useful life than the energy required to produce it.

BIBLIOGRAPHY

ANDERSON, L.L. 1972. Energy potential from organic wastes. Circular *8549,* Bureau of Mines, Washington, D.C.

APPELL, H.R., WENDER, I. and MILLER, R.D. 1970. Conversion of urban refuse to oil. Tech. Prog. Rep. *25,* Bureau of Mines, Washington, D.C.

BURWELL, C.C. 1978. Solar biomass energy: An overview of U.S. potential. Science *199,* 1041–1048.

BUXTON, J.W., WALKER, J.N. and COLLINS, L. 1977. Crop response in solar heated greenhouses ventilated with deep coal mine air. Proc. Conf. on Solar Energy for Heating Greenhouses and Greenhouse-Residential Combinations, March 20–23, Cleveland and Wooster, Ohio.

CALVIN, M., 1974. Solar energy by photosynthesis. Science *184,* 375–381.

CALVIN, M., 1977. A gasoline tree plantation? Agric. Eng. *58* (11) 12–16.

CLEWS, H. 1974. Electric Power From the Wind. Solar Wind, East Holden, Maine.

DANIELS, F. 1975. Direct Use of the Sun's Energy. Ballantine Books, New York.

DUFFIE, J.A. and BECKMAN, W.A. 1974. Solar Energy Thermal Processes. John Wiley & Sons, New York.

ENERGY RES. DEV. ADMIN. (ERDA). 1976. An economic analysis of solar water and space heating. *DSE-2322-1*. Energy Res. Dev. Admin., Washington, D.C.

GENERAL ELECTRIC COMPANY. 1977. Wind energy mission analysis. Prepared for Energy Res. Dev. Admin. under contract No. *EY-76-C-02-2578*. General Electric Space Div., Philadelphia.

GREELEY, R.S. and SPEWAK, P.C. 1976. Land and fresh water farming. Conf. on Capturing the Sun Through Bioconversion. Washington, D.C.

HAMMOND, A.L., METZ, W.D. and MAUGH, T.H. 1973. Energy and the Future. Amer. Assoc. for Adv. of Sci., Washington, D.C.

HARTSOCK, J.G. 1978. Solar grain drying conference proceedings. U.S. Dep. of Energy and U.S. Dep. Agric., Washington, D.C.

KIRIK, M. 1978. A canadian views alcohol as a farm fuel. Agric. Eng. *59* (5) 12–14.

MCCLOSKEY, M. 1978. An environmentist speaks: where should energy come from in the years immediately ahead? Energy (1978 Spec. Is.). 15–17. Business Communications Co., Stamford, Conn.

MCRAE, A., DUDAS, J.L. and ROWLAND, H. 1977. The sources of energy. *In* Energy Source Book. Aspen Systems Corporation, Germantown, Maryland.

NAT. SCI. FOUND. (NSF). 1976. Solar energy applications in agriculture: Potential, research needs and adoption strategies. *PB-265105*. Nat'l. Sci. Found., Washington, D.C.

POLLARD, W.G. 1976. The long-range prospects for solar-derived fuels. Amer. Sci. *64* (5) 509–513.

ROCKS, L. and RUNYON, R.P. 1972. The Energy Crisis. Crown Publishers, New York.

SMITH, C.C. 1978. Proc. Third Ann. Conf. on Solar Energy for Heating of Greenhouses and Greenhouse-Residence Combinations. U.S. Dep. of Energy and U.S. Dep. Agric. Washington, D.C.

SODERHOLM, L.H. and ANDREWS, J.F. 1974. Wind electric power. ASAE Paper No. *74-3503*. Pres. at 1974 Winter Meet. Amer. Soc. Agric. Eng., St. Joseph, Michigan.

STOUT, B.A. *et al.* 1978. Brazil promotes proalcool for petroleum independence. Agric. Eng. *59* (4) 30–33.

U.S. DEP. AGRIC. (USDA). 1977. Solar crop drying conference proceedings, Agric. Res. Serv.—U.S. Dep. Agric., Washington, D.C.

WILKE, C.R. 1977. Cellulose, food and energy. Report No. *5275*, Lawrence Berkley Laboratory, Univ. of Calif., Berkley.

10

The Future

Increasing world human population and predictions of a world population of 6.5 to 8 billion by 2000 are well known and generally accepted as inevitable. Food production increases commensurate with the expected population increases will be required to maintain the present less-than-adequate nutritional levels. Resources to produce this food are limited. Human population must at some point be controlled if the future is to offer much hope to the world's inhabitants. We can be certain that the population will be controlled although likely not as most desire. Progress will be made toward voluntary population control, but malnutrition and starvation, disease and war will each play too important a role.

To produce food for the world's increasing human population will require many more agricultural energy inputs than are now required. It is unknown and difficult to predict, however, whether per capita energy requirements for food production will increase or decrease in the future. There are powerful factors working in both directions, some tending to cause increases and some tending to cause decreases. We can only attempt to identify rather than quantify the effects of these factors.

Among the factors causing increasing per capita energy requirements for food production is the law of diminishing returns, defined here as decreasing partial and marginal energy productivities. This occurrence generally is true for many but not all increases of all inputs. Another factor is the result of increasing population with essentially constant arable land resources: decreasing arable land per capita, which leads to increased quantities of energy-requiring inputs in order to increase land productivity. A third factor is that, as resources are depleted, energy requirements for providing additional increments of the resources are increased (Chapman 1976). New nonrenewable resources do not come as easily as those already used.

On the other hand we also anticipate several factors operating to decrease per capita food energy requirements. Technological advances are certain to occur, many of which will result in greater energy productivity. Further increased energy productivity is expected through substitution of inputs, conservation efforts, and reduction of waste and spoilage. Enterprise shifts will lead to less energy input required per unit of food produced. We anticipate that decreases in agricultural energy inputs will occur due to changes in the character of animal agriculture, although likely without great decreases in the importance of animal agriculture. Also, agriculture is likely to produce a small but significant portion of the energy it consumes by utilizing renewable energy sources.

In the lesser developed countries (LDCs) of the world, agricultural energy inputs are expected to increase. It is the LDCs which have the greater portion of the world's population (3.2 billion) which is increasing faster (2.4% annually) and which is less well fed. Increased energy inputs to agriculture generally will produce the greatest yield increases when applied to the low energy agriculture which predominates in the LDCs, rather than in the already industrialized agriculture of the developed countries (DCs). Evidence of an increased importance and emphasis on agriculture by the LDCs would be welcomed; many LDCs have the potential to become self-sufficient in food production if more efforts were to be directed toward agricultural development instead of toward increased industrial, military and general economic development.

The agriculture of the LDCs will likely become more energy intensive through increases in fertilizer consumption, pesticide consumption, irrigation, yield-intensive mechanization and other inputs. The Green Revolution has set the stage in many LDCs for subsistence farmers to become cash or barter farmers and to begin to use energy-requiring inputs to increase their production. Support for the necessity of LDC farmers to increase their energy-requiring inputs is widespread. Revelle (1976) said that a considerable increase in energy use will be essential, primarily for irrigation, fertilizers and additional mechanical power. Makhijani (1975) agreed that irrigation and fertilizer use must increase and that machines and improved draft animals are high priority for many areas. Fertilizer increases were particularly emphasized by Smerdon (1977) and energy increases in general were spoken for by Stout and Myers (1977), Timmer (1975) and Crosson (1976).

If major new sources of energy are not developed, the food systems of the world will in the future likely consume a greater portion of the world's energy since food will remain as the most important output from man's efforts. This will leave less energy available to other industries probably causing them to diminish in relative importance. Resources, including labor, will be diverted from other industries to agriculture and the food chain; thus, it may be useful to anticipate such a situation in conjunction with further mechanization which will no doubt continue.

If major new sources of energy are developed such as advanced nuclear or solar technology, agriculture may experience major changes. The development of a hydrogen technology might then be feasible with plenty of electrical power for hydrolysis of water. Fertilizers and other inputs could then be derived from hydrogen.

It is perhaps only remotely possible that new sources of food will make significant contributions to the world's food requirements. These may emerge due to breakthroughs in technology or production capability associated with seafoods, direct conversion of hydrocarbons to foods, or space colonization. The authors tend to think, however, that conventional agriculture will continue to produce most of the food for man on earth.

BIBLIOGRAPHY

ANON. 1976. Energy for agriculture in the developing countries. Monthly Bull. Agric. Econ. and Statist. *25* (2) 1–8.

CHANCELLOR, W.J. 1977. The role of fuel and electrical energy in increasing production from traditionally based agriculture. East-West Center, Honolulu.

CHANCELLOR, W.J. and GOSS, J.R. 1976. Balancing energy and food production, 1975–2000. Science *192,* 213–218.

CHAPMAN, P.F. 1976. Principles of energy analysis. *In* Aspects of Energy Conversion. J.M. Blair, B.D. Jones, and A.J. Van Horn (Editors). Pergammon Press, New York.

CROSSON, P. 1976. Interdependencies between food sector and non-food sector activities with respect to energy resources and environment. Nat. Acad. Sci., World Food and Nutrition Study, Sub-group 10B, Washington, D.C.

FAIDLEY, L.W. 1977. Energy requirements and efficiency for crop production. ASAE Paper No. *77-5528.* Pres. at 1977 Winter Meet., Amer. Soc. Agric. Eng., St. Joseph, Michigan.

GREEN, M.B. 1978. Eating Oil. Westview Press, Boulder, Colorado.

KLEIS, R.W. *et al.* 1976. Energy related impacts on great plains agricultural productivity in the next quarter century, 1976–2000. Great Plains Agric. Counc., Univ. of Nebraska, Lincoln.

LIPTON, M. 1975. Energy and agriculture in poor countries. Span *18* (1) 17–18.

MAKHIJANI, A. 1975. Energy and Agriculture in the Third World. Ballinger Publishing Co., Cambridge, Mass.

PIMENTEL, D. *et al.* 1975. Energy and land constraints in food protein production. Science *190,* 754–761.

REVELLE, R. 1976. Energy use in rural India. Science *192,* 969–975.

SINGH, G. and CHANCELLOR, W. 1975. Energy inputs and agricultural production under various regimes of mechanization in northern India. Trans. ASAE *18* (2) 252–259.

SMERDON, E.T. 1977. The role of energy in food production—A perspective. *In* Energy Use Management. R. Fazzolare and C.B. Smith (Editors). Pergammon Press, New York.

STOUT, B.A. and MYERS, C.A. 1977. An overview of energy in developing countries. ASAE Paper No. *77-5527.* Pres. at 1977 Winter Meet. Amer. Soc. Agric. Eng., St. Joseph, Michigan.

TERHUNE, E.C. 1977. Prospects for increasing food production in less developed countries through efficient energy utilization. *In* Agriculture and Energy. W. Lockeretz (Editor). Academic Press, New York.

TIMMER, C.P. 1975. Interaction of energy and food prices in less developed countries. Amer. J. Agric. Econ. *57* (2) 219–224.

Appendix

APPENDIX A.1. GLOSSARY FOR AGRICULTURAL ENERGETICS

The authors have attempted to gather here definitions pertinent to agricultural energetics. There is not total agreement among the sources the authors of this book found; where there is more than one definition for a term, the one the authors preferred is given first. Also, there is a proliferation of terms having the same meaning as sequestered energy for agricultural products: embodied energy, gross energy requirement, energy requirement, cultural energy, energy subsidy and added energy. Sources are given for those definitions for which they can be identified.

Added energy: Total amount of energy used for the production of a product and the production of all inputs from the raw materials used, inclusive of the energy used for transportation (DeWit 1975; Anon. 1976).

Agricultural energy: Energy used in agriculture or produced through agriculture.

Ancillary energy: That energy required to produce inputs to agricultural production consumed in a single activity or production period, for example, the energy sequestered in fertilizers, chemicals and packaging materials; a subclassification of indirect energy (Gopalakrishnan *et al.* 1976).

Caloric gain: Ratio of food energy harvested to cultural energy consumed; inverse of Hirst's (1974) energy ratio (Heichel 1973).

Cultural energy: Sequestered energy for agricultural production, excluding solar (Heichel 1973).

Direct energy: Enthalpy of fuels used directly in a process under study plus the electrical energy delivered to the process (Hannon 1973; IFIAS 1974).

Embodied energy: The sum of all the direct and indirect energy required to produce a good or provide a service (Bullard and Herendeen 1975).

Gopalakrishnan *et al.* (1976) provided a more restrictive definition of embodied energy: that energy required to manufacture durable inputs such as tractors, machinery, buildings, refrigeration, equipment, etc.

Energy efficiency (also known as *energy ratio* or *energy efficiency ratio*): Quotient of energy value of (agricultural) output and energy value of the sum of all direct and indirect inputs (Pimentel *et al.* 1974).

Energy farming: The growing of crops to provide energy.

Energy intensity: Sequestered energy per unit of output, i.e., $MJ \cdot \$^{-1}$, $MJ \cdot t^{-1}$, etc. (IFIAS 1974, 1975).

Energy intensiveness (EI): $MJ \cdot t^{-1} \cdot km^{-1}$ for freight traffic or $MJ \cdot passenger^{-1} \cdot km^{-1}$ for passenger traffic.

Energy intensiveness: Amount of energy needed per unit of crop produced (Commoner *et al.* 1975).

Energy investment ratio (EIR): Ratio of cultural energy to natural energy, or, in a more general context, ratio of feedback energy to a secondary energy source, both in FFEs (Odum and Odum 1976).

Energy productivity: Quotient of agricultural product produced and energy sequestered in its production. The reciprocal of energy intensity.

Energy quality factor (EQF): Ratio of total energy inputs to energy outputs, both in FFEs.

Energy subsidy: Sequestered energy for agricultural production (Slesser 1973).

Energy yield ratio: Energy output from an energy source divided by feedback energy required to produce it, both in FFEs (Odum and Odum 1976).

Enthalpy (H): The sum of molecular and flow energy ($H = U + PV$). Gross heat of combustion of fuel.

Entropy (S): Energy unavailable for work or bound energy. Low entropy is a measure of the ability of a system to undergo spontaneous change and a measure of high energy quality. High or increasing entropy is a measure of increasing randomness or disorder. Change in entropy $dS = dQ/T$ with $Q =$ energy and $T =$ absolute temperature of conversion.

Feedback energy: Energy returned from a storage to influence the process which places it in storage. The energy required to get more energy (Odum and Odum 1976).

Fossil fuel equivalents (FFE): Factor by which to multiply the energy of any quality to convert heat equivalents to the energy equivalent of a fossil fuel. The number of units of energy of a fossil fuel equivalent to a unit of any other form of energy. High quality energy has high FFE (Odum and Odum 1976).

Free energy: Thermodynamic potential expressed by Gibbs' function $G = H - TS$ where $H =$ enthalpy, $T =$ temperature and $S =$ entropy. H is about 10% more than G for most intensive fuels.

Gross energy requirements (GER): Amount of energy source sequestered by the process of making a good or service; includes feedback energy to obtain energy output (IFIAS 1974).

Indirect energy: Energy used by suppliers of materials and services to an industry and by suppliers of these suppliers, etc. (Hannon 1973).

Marginal energy productivity: The quotient of the additional agricultural product resulting from an additional unit of energy sequestered in one or more inputs.

Net energy (also *surplus energy*): Energy output from an energy source in excess of the feedback energy required to produce it (Odum and Odum 1976; Cottrell 1955).

Net energy analysis (NEA): Analysis of energy supply and conservation technologies which compares the energy output to the feedback energy to obtain the supply or conserve a flow, both measured in heat units.

Net energy requirements (NER): Gross energy requirement less the gross enthalpy of combustion of the products of a process (IFIAS 1974).

Partial energy productivity: The quotient of the additional quantity of agricultural product resulting from the addition of any quantity of a single production input and the energy sequestered in that input.

Primary energy: Fossil fuel energy extracted from the earth, expressed as total enthalpy (Bullard and Herendeen 1975).

Process energy requirements (PER): Direct and indirect energy delivered to a process. PER≤GER (IFIAS 1974).

Sequestered energy: The sum of all the direct and indirect energy required to produce a good or provide a service (IFIAS 1974).

TABLE A.2. ENERGY EQUIVALENTS[1]

	bbl oil[2]	Btu	ft · lb	hp · h	MJ	kcal[3]	kWh
bbl oil[2]	1	5.7×10^{6}	4.4×10^{9}	2.2×10^{3}	6.0×10^{3}	1.4×10^{6}	1.7×10^{3}
Btu	1.8×10^{-7}	1	778.0	3.929×10^{-4}	1.055×10^{-3}	0.2520	2.931×10^{-4}
ft lb	2.3×10^{-10}	1.286×10^{-3}	1	5.051×10^{-7}	1.356×10^{-6}	3.238×10^{-4}	3.766×10^{-7}
HP-hr	4.5×10^{-4}	2545	1.980×10^{6}	1	2.685	641.2	0.7457
MJ	1.7×10^{-4}	947.8	0.7376×10^{6}	0.3725	1	238.8	0.2778
kcal[3]	7.0×10^{-7}	3.968	3088	1.560×10^{-3}	4.187×10^{-3}	1	1.163×10^{-3}
kWh	6.0×10^{-4}	3412	2.655×10^{6}	1.341	3.600	860.0	1

[1]To convert from a unit on the left to a unit in a column, multiply the known quantity by the factor in the table; i.e., 2500 kcal $\times$ 1.163×10^{-3} kWh/kcal = 2.908 kWh.

[2]Approximate conversion only; varies with source of crude oil. 42 gal.

[3]1 kcal (large calorie or kilogram calorie) = 1000 cal (small calorie or gram calorie)

BIBLIOGRAPHY

ANON. 1976. Energy for agriculture in the developing countries. Mon. Bull. Agric. Econ. and Stat. *25* (2) 1–8.

BULLARD, C.W. and HERENDEEN, R.A., III. 1975. The energy cost of goods and services. Energy Policy *3* (4) 268–278.

COMMONER, B., GERTLER, M., KLEPPER, R. and LOCKERETZ, W. 1975. Energy Problems. *CBNS-AE-2,* Cntr. for the Biol. of Nat. Sys., Washington Univ., St. Louis.

COTTRELL, F. 1955. Energy and Society. McGraw-Hill, New York.

DEWIT, C.T. 1975. Substitution of labour and energy in agriculture and options for growth. Neth. J. Agric. Sci. *23,* 145–162.

GOPALAKRISHNAN, C., PATRICK, N.A. and CHANCELLOR, W.J. 1976. Determining energy requirements for agriculture: some methodological issues. Ppr. presented at Amer. Agric. Econ. Assoc. Meeting, Aug. 15–18. Penn. State Univ., University Park.

HANNON, B.M. 1973. An energy standard of value. J. Ann. Amer. Acad. Pol. Soc. Sci. *410,* 139–153.

HEICHEL, G.H. 1973. Comparative efficiency of energy use in crop production. Bull. *739,* Conn. Agric. Exp. Sta., New Haven.

HIRST, E. 1974. Food-related energy requirements. Science *184,* 134–138.

INT. FED. INST. ADV. STUD. (IFIAS). 1974. Energy Analysis Workshop Rept. No. *6.* Int. Fed. of Inst. for Adv. Stud., Stockholm, Sweden.

IFIAS. 1975. Workshop on Energy Analysis and Economics. Report No. *9.* Int. Fed. of Inst. for Adv. Stud., Stockholm, Sweden.

ODUM, H.T. and ODUM, E.C. 1976. Energy Basis for Man and Nature. McGraw-Hill, New York.

PIMENTEL, D. *et al.* 1974. Workshop on Research Methodologies for Studies of Energy, Food, Man and Environment, Phase I. Cornell Univ., Ithaca, New York.

SLESSER, M. 1973. Energy subsidy as a criterion in food policy planning. J. Sci. Food Agric. *24,* 1193–1207.

STEINHART, J.S. and STEINHART, C.E. 1974. Energy use in the U.S. food system. Science *184,* 307–316.

Index

Other AVI Books

AGRICULTURAL AND FOOD CHEMISTRY
Teranishi
AGRICULTURAL PROCESS ENGINEERING
Henderson, Perry 3rd Edition
AN INTRODUCTION TO AGRICULTURAL ENGINEERING
Roth, Crow, Mahoney
BREEDING FIELD CROPS
Poehlman 2nd Edition
FOOD AND ECONOMICS
Hungate, Sherman
FOOD ENGINEERING SYSTEMS
Farrall Vol. 1 and 2
FOOD MICROBIOLOGY: PUBLIC HEALTH AND SPOILAGE ASPECTS
deFigueiredo, Splittstoesser
FOOD QUALITY ASSURANCE
Gould
FOOD SANITATION
Guthrie
FUNDAMENTALS OF FOOD ENGINEERING
Charm 3rd Edition
NUTRITIONAL EVALUATION OF FOOD PROCESSING
Karmas 2nd Edition
NUTRITIONAL QUALITY INDEX OF FOODS
Hausen, Sorenson
PACKAGING REGULATIONS
Sacharow
PLANT PHYSIOLOGY IN RELATION TO HORTICULTURE
Bleasdale
PRINCIPLES OF ANIMAL ENVIRONMENT
Esmay Textbook Edition
PROCESSING EQUIPMENT FOR AGRICULTURAL PRODUCTS
Hall, Davis 2nd Edition
PROGRESS IN HUMAN NUTRITION
Margen, Orgar Vol. 2

QUALITY CONTROL FOR THE FOOD INDUSTRY
Kramer, Twigg *3rd Edition* *Vol. 1 and 2*
SOILS AND OTHER GROWTH MEDIA
Flegmann, George
SOIL, WATER AND CROP PRODUCTION
Thorne, Thorne
SYSTEMS ANALYSIS FOR THE FOOD INDUSTRY
Kramer